UNDERSTANDING LICHENS

George Baron

The Richmond Publishing Co. Ltd. would like to thank
Frank S. Dobson, Jeremy Gray and David Hawksworth CBE for all their
assistance in publishing this title.

Original text and illustrations © 1999 George Baron
Figs. 8, 11, 16, 19 (part), 24, 29, 31, 32 and 33 © 1999 Ewart Thomas
Additional text and illustrations © 1999 Frank S. Dobson and Jeremy Gray
'Churchyard lichens' © 1999 T. Chester
'Woodland and parkland lichens' © 1999 Francis Rose

ISBN 085546 252 3

The Richmond Publishing Co. Ltd.
P.O.Box 963 Slough SL2 3RS England

Telephone : +44 (0) 1753 643104
Fax : +44 (0) 1753 646553
email rpc@richmond.co.uk

Front cover photographs © 1999 Jeremy Gray

Background picture: Manorbier coast, Pembrokeshire
From left to right: *Cladonia floerkeana*; *Caloplaca flavescens*; *Physcia leptalea*;
Solorina saccata.

All other photographs © 1999 Frank S. Dobson

Preface

The South London Botanical Institute was founded in 1910 by a distinguished Indian Civil Servant, Allan Octavian Hume, who was remarkably the founder of the Indian Congress Party and its General Secretary for the first twenty-two years of its existence. He ultimately retired to the Crystal Palace area and set up the Institute with the intention of 'promoting, encouraging and facilitating among the residents of South London the study of the science of Botany'. The Institute has, with its well tended botanic garden, its herbaria (which includes a substantial lichen section), its organized classes and rambles and other activities, carried out the founder's purposes with surprising consistency over the years. It is maintained by endowments left by its founder, by members' subscriptions and donations, and by the proceeds of plant sales and other events. For its day to day operation it depends entirely on voluntary work by its members.

This publication is the first of its kind produced by the Institute. I am most grateful to several of my co-members for their help. Ewart Thomas has provided the illustrations (supplemented with illustrations from *Lichens: An Illustrated Guide* by Frank S. Dobson) and has made valuable suggestions for the text. Graeme Lyall, Warden of the Institute, designed the first (desk-top) edition and thus enabled the venture to be launched

I also wish to express my warm thanks to Frank Dobson and Jeremy Gray for all the time and thought they have given to the revision of the text and for supplying the colour illustrations which add so much to this second edition; to Tom Chester and Francis Rose for supplying notes on churchyard and woodland lichens; to Peter James, David Hawksworth, Ivan Pedley, David Streeter and Pat Wolseley for their valuable suggestions. Finally, to Judy Marshall, Chair of the Council of the Institute, has given her full support throughout.

Membership of the Institute is open to all who wish to pursue an interest in plants. For further details, please write to the Membership Secretary, SLBI, 323 Norwood Road, London SE24 9AQ.

George Baron
September 1999

Contents

Introduction

The purpose of this book is to arouse in the reader an interest in lichens, one of the 'lesser' and certainly lesser-known, life-forms. Indeed, very many people may have never noticed lichens at all, since they often appear as no more than unobtrusive patches of colour on tree-trunks, rocks and walls.

Awareness of, and interest in, the natural world has increased and is increasing in this 'Green' period and is directed predictably towards mammals, birds, fishes, trees and flowers which have an immediate appeal to the emotions and the senses, especially if rare or threatened. The more inconspicuous forms of life, such as mosses, ferns, seaweeds and grasses are not even given the same value as during the Victorian era and lichens seem valued almost least of all. The writer of one of the first popular account of lichens, William Lauder Lindsay, displays a touching enthusiasm in claiming attention to their charms. He writes, in 1856:

'The delicate waving frond of the fern is anxiously tended by jewelled fingers in the drawing-rooms of the wealthy and noble; the rhodospermous seaweed finds a place beside the choicest productions of art in the gilt and broidered album; the tiny moss has been the theme of many a gifted poet; and even the despised mushroom has called for the classic works in its praise. But the Lichens, which stain every rock and clothe every tree, which form:

> *'Nature's livery o'er the globe*
> *Where'er her wonders range,'*

have been almost universally neglected, nay despised.'

He goes on to explain that little is needed for the study of lichens:

'The lichenological student requires no cumbrous or expensive apparatus; an old knife and hammer, a few pill-boxes or a tin-can for collecting, a supply of cardboard and paper, with gum or glue for preserving, and a pocket-lens and microscope for examining, constitute his whole armamentaria.'

In its simplest form, the study of lichens through their collection and identification, makes no more demands on the beginner today, although he or she would now certainly be equipped with small quantities of the chemicals which have come into use since the time of Lindsay.

As the passages from Lindsay indicate, the nineteenth century amateur naturalist was, above all, a collector. He freely plundered all living creation with none of the inhibitions which now restrain his successors. The great museums and herbaria remain as monuments to his energy and the risks he often took in far corners of the world.

Things are different now. Collecting, even with discrimination, is rightly seen as a potential threat to our fragile environment, except as a means of saving rare species from extinction. In its place there has developed an increasingly sophisticated interest in nature photography both as still images or in the vivid sequences of film and television.

A consequence is that an imperceptible constraint has developed between us and an active study of the natural world. The sheer magnificence of 'media' presentations makes possible a readily accessible but often rapidly forgotten sequence of visions of fascinating animals, insects and plants of all kinds. But it does not in itself lead to any depth of knowledge or understanding. Moreover, the increasingly exacting degree of knowledge made by scientific literature, especially of that dealing with lichens and other lesser known life-forms, can soon kill off any developing interest that the amateur may experience.

What this book attempts to do is to present, in as simple a form as possible, a basic understanding of the nature of lichens, their forms, their physiology and their habitats. It is hoped that, as a result, its readers may be encouraged to feel enough confidence to undertake explorations both of the study of lichens in the field and also of the literature of lichenology which is surveyed in a final chapter.

The reader will be introduced to many new terms early on in this book. All of these are explained under the entry in the index.

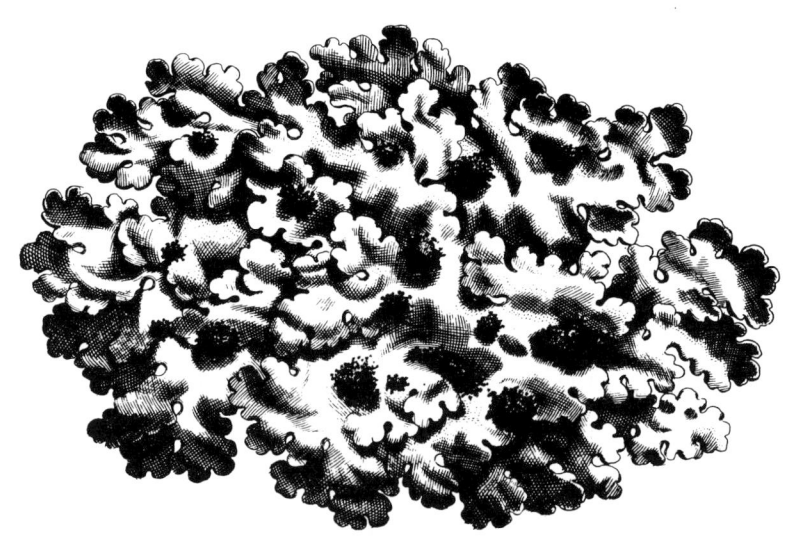

A foliose lichen from Micheli 'Nova Plantarum Genera' 1727

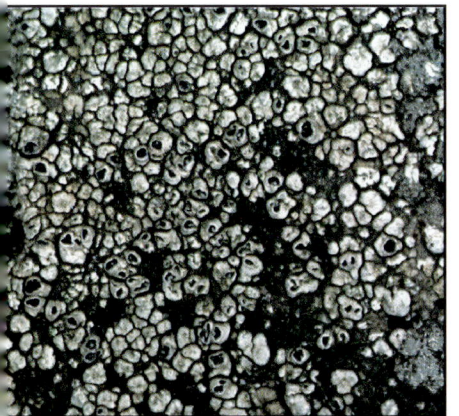

1. *Aspicilia contorta* x5

2. *Caloplaca citrina* x5

3. *Caloplaca flavescens* x3

4. *Candelariella medians* x2

5. *Cladonia portentosa* x3

6. *Cladonia pyxidata* x4

Plate 1

7

7. *Collema crispum* x3

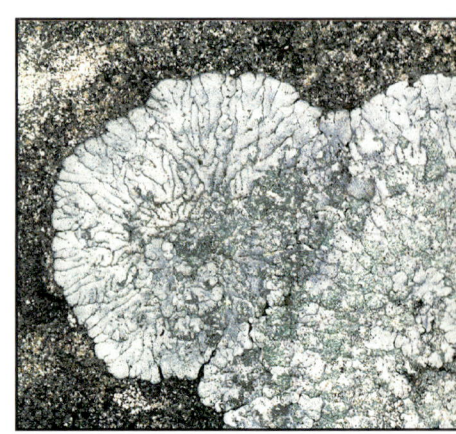

8. *Diploicia canescens* x4

9. *Graphis elegans* x10

10. *Lasallia pustulata* x3

11. *Lecanora conizaeoides* x6

Plate 2

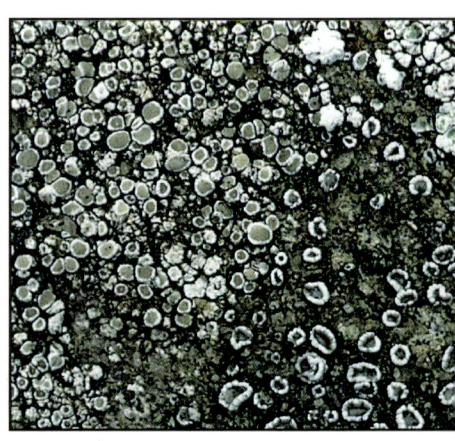

12. *Lecanora dispersa* x10

8

13. *Lepraria incana* x4

14. *Lobaria pulmonaria* x0.75

15. *Ochrolechia parella* x3

16. *Parmelia subrudecta* x2

17. *Peltigera membranacea* x2

Plate 3

18. *Physcia adscendens* x6

9

19. *Physcia caesia* x4

20. *Ramalina fastigiata* x3

21. *Rhizocarpon geographicum* x6

22. *Usnea florida* x3

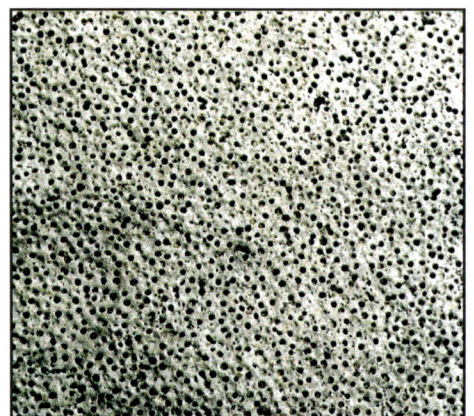

23. *Verrucaria baldensis* x8

Plate 4

24. *Xanthoria parietina* x2

Plate 1.

1. *Aspicilia contorta*: A common crustose species that is found on hard calcareous rocks and memorials. It forms small, separate areoles when young. Many of the areoles contain a single, innate, black centred apothecium. With age, the areoles merge to form a continuous crust.

2. *Caloplaca citrina*: Found on concrete and other calcareous substrata where it is very common. It is often infertile when it consists of very small, lemon yellow to orange granules.

3. *Caloplaca flavescens*: Common on calcareous substrata, especially limestone memorials. It is crustose with a placodioid margin. The fruits are bright orange and found towards the centre of the thallus. The white area just inside the lobes is commonly found in this species.

4. *Candelariella medians*: A rare species on natural limestone but now very common on limestone memorials in semi-polluted areas .

5. *Cladonia portentosa*: One of the 'reindeer lichens' and is found on dunes and heathland. The granular primary thallus soon dies away leaving the podetia tangled amongst the vegetation.

6. *Cladonia pyxidata*: A cladoniform lichen found on soil on walls, mossy rocks and trees. The illustration shows the squamulose primary thallus and the vase-shaped podetia with corticate granules inside the cup.

Plate 2.

7. *Collema crispum*: A gelatinous, homoiomerous, lichen found on calcareous rocks and mortar. It has globose to flattened isidia on the lobes.

8. *Diploicia canescens*: A placodioid lichen that is very common on nutrient-enriched rocks and trees. It is white-pruinose with white to grey soredia. Very rarely fertile.

9. *Graphis elegans*: A lirellate lichen. Found on smooth-barked trees in reasonably clean air. The hard margins of the apothecia are scored with very fine grooves.

10. *Lasallia pustulata*: 'Rock tripe'. An umbilicate lichen, attached only by a central holdfast. Found on nutrient-enriched acid rocks, mainly in upland regions.

11. *Lecanora conizaeoides*: Very pollution resistant. In polluted regions it often completely covers trees, fences and acid rocks. It is becoming less common as pollution levels fall. The apothecia usually have granular margins.

12. *Lecanora dispersa*: Like the previous species it is common in polluted areas but is also found on almost any concrete surface or limestone throughout the British Isles.

Plate 3

13. *Lepraria incana*: A common leprose species that forms a loose crust of minute, fluffy granules. It is found in shaded sites such as sheltered racks and tree bases.

14. *Lobaria pulmonaria*: 'Lung wort'. A large foliose species found in clean air regions especially in ancient woodlands.

15. *Ochrolechia parella*: A crustose species of acid rocks. It has wide margined fruits with granular pruina on the disc. The margin often shows growth-rings.

16. *Parmelia subrudecta*: A common foliose lichen. It is found on trees and mossy rocks. Like all *Parmelia* species it is attached to the substratum by root-like rhizinae. It has small, pale, point-like soralia on the middle of the lobes and less frequently along the margins. On the left in this illustration is *Parmelia revoluta*. This has smooth, buff-tipped, bluish-grey lobes. When mature, granular soredia are found along the tips of the lobes.

17. *Peltigera membranacea*: The 'dog-lichen', so named from the tooth-like rhizinae. A common leafy lichen found in damp sites on mossy rocks, trees and even mown lawns.

18. *Physcia adscendens*: Very common on calcareous substrata. The narrow lobes become almost upright and the swollen tips burst to expose the soredia inside. The lobes also have white or piebald cilia along their margins.

Plate 4.

19. *Physcia caesia*: A foliose species but closely adpressed to the substratum. It is common on nutrient-enriched, well-lit calcareous materials but may also be found on acid rocks on the coast. The centre of the thallus and the long lobes have pale, blue-flecked soredia.

20. *Ramalina fastigiata*: A fruticose species that is found on nutrient-enriched, well-lit trees and hedges. Rarer in upland regions. The flattened lobes have pale-centred apothecia on the tips.

21. *Rhizocarpon geographicum*: The 'map lichen'. A distinctive and common crustose species of upland and coastal acid, well-lit rocks. It is variable in colour from bright yellow to almost green when found in light shade.

22. *Usnea florida*: A fruticose species found in the upper canopy of trees in unpolluted regions. Like all *Usnea* species, it has a central holdfast and a central core that becomes visible if you stretch a thallus until it splits. The large disc-shaped apothecia are surrounded by long fibrils.

23. *Verrucaria baldensis*: A common species of hard calcareous rocks and memorials. The thallus is almost entirely within the rock with only the dark prothallus showing the limit of the thallus. The fruits are in in small holes with only the tips of the perithecia visible under a hand-lens. In this species, the normally hard, flask-shaped perithecium only has a 'trap-door' covering the fruit. This means that when a perithecium dies it leaves an empty hole in the substratum.

24. *Xanthoria parietina*: A very common foliose lichen. It is found on nutrient-enriched sites from old bones to farmyards. It is especially common on well-lit roofs under bird-perching sites such as television aerials.

Chapter 1
What is a Lichen?

Lichens are clearly different from the flowering plants that grow in a garden or by the wayside. Lichens do not have flowers, seeds or true roots and the whole structure is much simpler. It may be difficult to separate them, visually, from the bryophytes (mosses and liverworts), which have true leaves. Like lichens, these plants also reproduce by means of spores, but the presence of a round, tubular or umbrella-shaped spore-case on a stalk of a centimetre or more, in itself generally distinguishes them from lichens.

The lichen symbiosis

Though there is still much debate over the answer to the question 'What is a lichen?', it is more or less agreed that it is a fungus (the mycobiont) together with one or more algae and/or cyanobacteria (the photobiont) living together in a symbiotic relationship and forming a stable, identifiable body. Even here there is uncertainty over how far it is a mutualistic relationship or whether the photobiont is subjected to controlled parasitism by the mycobiont.

Symbiosis is the term used to describe the phenomenon of organisms sharing each other's life processes in some way, for the benefit of both. This may range from a situation in which one is dominant and parasitizes the other, to a situation in which both gain equally from the association.

In a parasitic relationship the dominant organism damages or even destroys its host. This happens in the case of the flies that lay their eggs in the larvae of other insects or in the living tissue of animals. Another example is the deadly ichneumon fly which preys on caterpillars. At the other end of the scale is the relationship between birds and the insects that they feed on which prey on buffalo and other grazing animals, and the bacteria which live in the digestive tracts of animals forming an essential element in the processes of digestion.

The symbiotic relationship, which is fundamental to lichens as a growth-form, was not understood by the first lichenologists. Early in the nineteenth century, scientists observed small green cells in the lichen thallus (the body of the lichen). These were thought, at first, to be reproductive organs and were termed 'gonidia' (Gr. *gonad*, germ-cell). But the resemblance between these gonidia and the cells of algae was eventually noted and the way was open for the work of the Swiss botanist Schwendener, who may justifiably be regarded as the discoverer of the true nature of lichens. In 1867 he published his findings on the relationships of fungal hyphae and gonidia in lichens. He argued that the latter were algal cells and that the hyphae were parasitic on them. His basic conclusions are now universally accepted, but subsequent thinking moved towards a more benevolent form of association, a mutualistic relationship in which both elements were partners in a joint struggle for survival in harsh conditions.

The use of the term 'controlled parasitism' marks a reassessment by those who judge that the balance of advantage lies with the fungal component.

The fungal partner is, in most cases, unable to live on its own and depends on sugars produced by the photobiont for its only nutriment. If cultured separately in the laboratory the fungus rarely develops beyond a shapeless mass of hyphae (the fungal filaments). This is not the case with the photobiont, which is able to grow in a free-living state producing the sugars that it needs by photosynthesis. Nevertheless, within the lichen thallus the photobiont is protected from environmental pressures by the intermeshed fungal hyphae. These serve, mainly, to form a firm supporting structure and hence the lichen's shape and general appearance. The protection afforded by the fungus enables the photobiont to grow in harsh situations where it would otherwise be impossible for it to grow on its own. It acts as a shield against excessive light, resisting desiccation and helping to provide and store water and mineral salts derived from the atmosphere and the substratum (the rock, tree, soil, etc. on which the lichen grows). However, it is the requirements of the photobionts which determine where the lichen can grow, but the composition of the substratum, the atmospheric conditions, the moisture level and light intensity are all vital to the survival and growth of both partners. This means that a balance has to be constantly maintained between the two partners. In most lichens the fungus forms 85–95% of the mass of the thallus. Therefore, as the lichen grows and new fungal hyphae are produced, the photobiont must form additional cells and position them in this new fungal tissue at a rate that maintains this balance and enables photosynthesis to take place in the most efficient manner.

In some lichens, the thallus is formed by the fungal hyphae being just wrapped around a filamentous alga, the lichen assuming the hair-like character of the alga. In gelatinous lichens the shape, colour and other characteristics are determined more by the photobiont, a cyanobacterium, which forms a proportionately larger part of the thallus.

The fruiting bodies produced by the lichen are vital in classification and it is important to note that these are produced solely by the fungus. The photobiont does not normally reproduce sexually. Since fungi that form lichens cannot grow and fruit in a free state they must find appropriate photobiont partners at an early stage of development if they are to survive. There is still little known about how spores contrive to come together with the right kind of photobiont and there is much conjecture among lichenologists. It certainly seems that the chances of ascospores finding free-living algae must be remote. It may be that some spores may lie dormant until stimulated by a nearby alga to germinate. The algae *Trebouxia* and *Pseudotrebouxia* were thought not to exist in a free-living state in nature but they have been found in close enough proximity to lichens such as *Xanthoria parietina* (pl. 4) to suggest incorporation. It has, however, been established in a number of species that if a spore alights on another lichen with an appropriate photobiont it can acquire its algal cells for its own purposes. In time it will parasitize the host lichen and eventually replace the host's fungal tissue with its own.

Within the lichen thallus the algal cells are usually arranged in a layer just below the upper surface or with cyanobacteria, often spread throughout the thallus (Figs. 1 and 2).

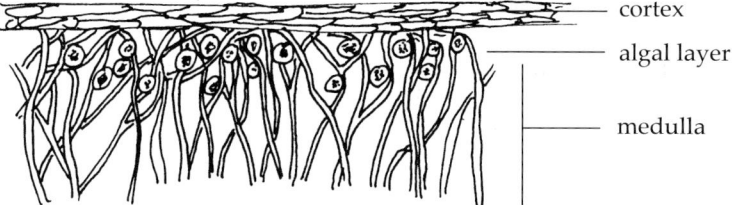

Fig. 1. Diagramatic section through a crustose lichen

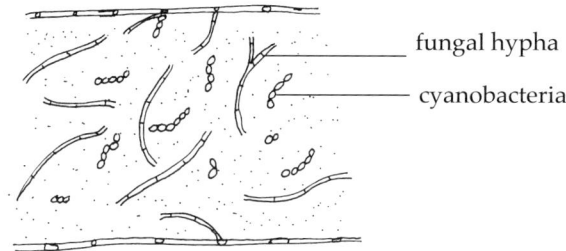

Fig. 2. Diagramatic section through a gelatinous lichen

Contact between mycobiont and photobiont takes place in several ways. There may be little discernible contact, or the photobiont cells may be lightly or deeply enmeshed in the fungal hyphae. Transfer of carbohydrates from photobiont to mycobiont is either through specialized hyphae 'appressoria' which press against the photobiont cells, or with 'haustoria' which penetrate their cells. Transfer is effected by diffusion through the cell walls of both partners. Substances produced by the fungus make the cell wall of the alga more permeable so that it may leak up to 80% of its sugars for use by the host.

The dual character of lichens

This was not understood until the middle of the nineteenth century. It is not surprising, therefore, that they were accorded their own taxonomic group and were not included in that of the fungi. Now, the classification integrates lichenized and non-lichenized fungi in a single scheme of classification. This is no easy task, since there is a long tradition of lichenology having its own specialized literature and herbaria and since they have been, and still are, neglected by most mycologists.

In the world as a whole there are, according to Ainsworth and Bisby in *Dictionary of the Fungi* (1995) some 525 genera and 13,500 species of lichens, but Hale, in *The Biology of the Lichens* (3rd edn 1983) puts the total number of species at about 17,000. The latter figure means that, on balance, about 20–25% of the 72,000 known fungi are lichenized. Amongst the ascomycetes, Hawksworth and Hill in *The Lichen-forming Fungi* (1984) consider that as many as 46% of species are lichenized. Although there are at least

1,500 genera of algae, only three provide 85% of the green photobionts found in lichens. *The Lichen Flora of Great Britain and Ireland* (1992) lists over 1,799 lichen species in 262 genera. Any such figures must, of course be very tentative as new discoveries are constantly being made. There are still many areas of the world which have not been adequately surveyed. This is particularly true of tropical and subtropical grasslands and forests. New species are frequently being found even in the well-populated areas of the world. Changes also frequently occur as a result of taxonomic revision.

Lichenologists tend to be either 'lumpers' or 'splitters', that is, those who choose to ignore minor variations and those who seek to create new species from them. Recent techniques such as DNA sequencing and cladistics are adding to a clearer understanding of the relationships between lichen species and as our knowledge develops there is much revision of the names of lichens.

Because of their fragility the study of the evolution of lichens is beset with many difficulties. Fossil remains are very few. The oldest specimens so far to be recorded date back to the Devonian era of approximately 400 million years ago. Others have been found, including *Hypogymnia physodes* in amber, dating back to the Caenozoic era of 25–70 million years ago. A recent discovery is that of specimens of *Peltigera aphthosa* in peat in the permafrost (now retreating) in Spitzbergen.

How lichens originated is a matter of conjecture. The suggestion has been made that they resulted from fungi, in harsh conditions, absorbing substances from neighbouring algae or cyanobacteria and then proceeded to develop a more permanent relationship. This could well have happened shortly after life first left the sea. It is now agreed that all lichens are 'polyphyletic', i.e. they are not derived from a single ancestor but that the process of lichenization has taken place many times in evolution. However it occurred, the relationship has proved very successful and lichens are found in many habitats where other organisms would find it very difficult to survive, from the icy wastes of Antarctica to the baking deserts of Saudi Arabia. It has even been suggested that lichens could possibly live in the harsh conditions on other planets.

The mycobiont is dominant in lichens in that it provides both their structure and their fruiting bodies, but it is the photobiont that is essential for their sustenance because of its capacity to carry out the process of photosynthesis, i.e. the absorption of light to form carbohydrates from the carbon dioxide and water in the atmosphere. Nevertheless, the total lichen resembles neither mycobiont nor photobiont and it is true to say that the whole lichen is greater than the sum of its individual parts.

The mycobiont

Well over 95% of the fungi that form lichens are members of the Ascomycota (producing their spores in a bag-shaped ascus, Fig. 3). Most of the remaining lichen-forming fungi are found in the Basidiomycota (producing their spores on the top of cushion-shaped basidia, Fig. 4) and these often grow mushroom-shaped fruiting bodies. There are also a few mycobionts from the Deuteromycota (the 'fungi imperfecti' which only reproduce asexually). There are also lichen-like relationships between other fungi such as the Myxomycetes (the slime moulds) and these are still being investigated. Some genera of fungi consist entirely of lichenized species whilst there are many others that contain both lichenized and non-lichenized species. About 20% of known species of fungus have adopted the lichen habit as the means of obtaining their nutrition.

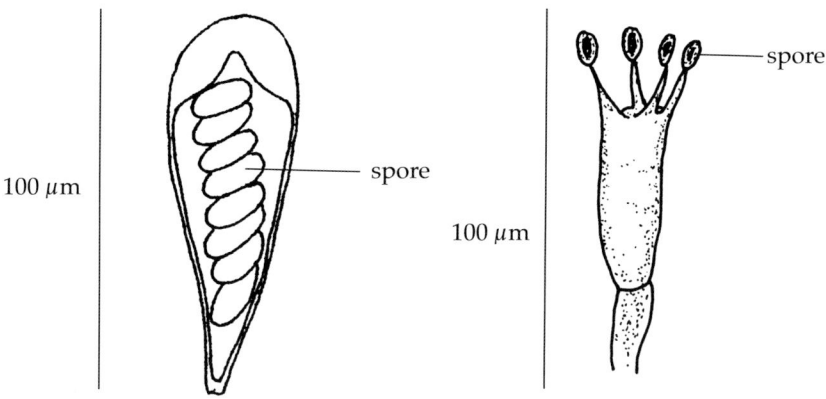

Fig. 3. An ascus   Fig. 4. A basidium

Every lichen is a different species of fungus but only a few genera of algae and cyanobacteria are incorporated. The name applied to the lichen is therefore only that of the fungus. From the taxonomic point of view, it is on the characteristics, expressed in thallus form, ascus structure, spores and other features that lichen classification is based.

The photobiont

Algae
These are simple plants that use chlorophyll to photosynthesize and thereby produce carbohydrates for nutrition. They include the seaweeds which can grow to many metres long. However, the algae incorporated in lichens are just small, simple or grouped cells barely visible to the naked eye. When these algae are free-living they form minute clusters or chains.

The algae found in lichens are mainly *Trebouxia* (Fig. 5), *Trentepohlia* (Fig. 6), *Myrmecia* and *Coccomyxa*, of which the first two are much the commonest. *Trebouxia* (including *Pseudotrebouxia* which is sometimes separated due to its manner of cell division to produce 'cell packets') comprises from half to three-quarters of all the species of algae found in lichens. The other common alga *Trentepohlia* forms branched chains of cells and contains an orange pigment. It is frequently found free-living on the damper shaded side of trees and rocks. It is therefore found incorporated mainly into shade-dwelling lichens. Its presence can often be discovered by scratching through the outer layer (the cortex) of the lichen where an orange streak will be seen. Although this alga appears orange to the naked eye, if viewed under a microscope the normal green chloroplasts are clearly visible.

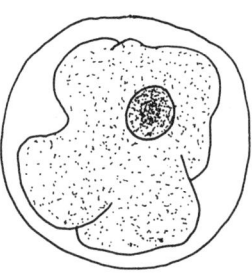

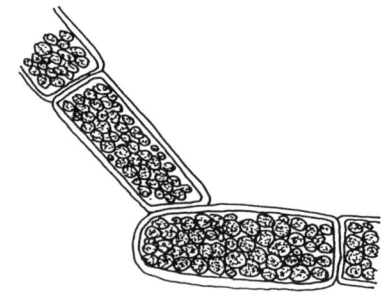

Fig. 5. *Trebouxia*   Fig. 6. *Trentepohlia*

Cyanobacteria
These differ from the green algae in a number of respects, especially in the lack of a membrane enclosing the nucleus. The cyanobacteria are unique in forming thick-walled cells called heterocysts which have the ability to fix nitrogen from the air. Common genera include *Nostoc* (Fig. 7), *Scytonema* and *Stigonema*, of which *Nostoc* is the most frequent, and it is also commonly found free-living, as an olive-brown gelatinous sheet. The production of heterocysts is greatly increased in *Nostoc* when it is incorporated in a lichen. Under the microscope these cyanobacteria have a blue-grey to brown tinge. It will be seen that they produce a thick gelatinous matrix in the lichen. The lichens of which they form part are often also brown or bluish-brown (e.g. *Collema* (pl. 2), *Leptogium*, *Peltigera*), especially when wet. Cyanobacteria are found in some species of lichen containing green alga, mostly in small, internal or external packets called 'cephalodia'. In many other lichens they are the sole photobiont component.

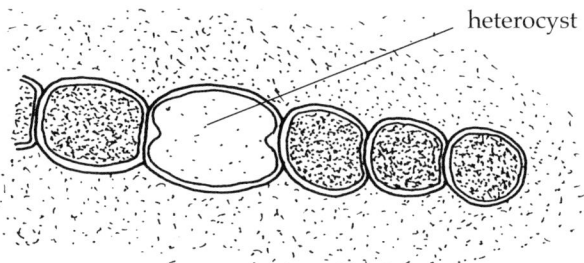

heterocyst

Fig. 7. *Nostoc* in gelatinous matrix

The identification of algae and cyanobacteria, certainly down to species level, is difficult, owing to the degree of deformity resulting from lichenization. Opinions vary even about the number of the genera found in lichens. A conservative estimate puts the total at 26 (out of a total of about 1,600 genera of photobiont worldwide), but other estimates are as high as 40 or more. This number may be compared with the more than 3,500 genera of lichens. Of all photobionts found in lichens about 90% are algae and only about 10% cyanobacteria.

It is surprising but true that a lichen species may accommodate different photobionts. *Collema* species are found only with *Nostoc*, but species of *Lichina* normally contain *Calothrix* or *Dichothrix* but may be found with other species of cyanobacteria capable of lichenization. Some species of *Sticta* and *Nephroma* can equally well contain either algae or cyanobacteria. The interaction of the substances produced by the two partners gives rise to two independent lichens of different appearance. They both, however, contain the same fungus and therefore can only have one name. These two forms are referred to as 'morphs' or 'phases' of that species. This has meant that species hitherto considered separate and distinct, but differentiated solely by the one having an alga and the other a cyanobacterium, have had to be renamed. Thus *Lobaria amplissima*, which is leaf-like, has green algae, but it may acquire a cyanobacterium, and develop rather brain-like cephalodia which can break away and form a different looking lichen which used to be called called *Dendriscaulon umhausense* but is now better described as a 'morph' of *Lobaria amplissima*. A similar occurrence of two very different forms of one lichen is also found in *Peltigera britannica*.

Chapter 2
The Lichen Thallus

It is, above all, the fungal element that gives a lichen its form and most of its characteristics. This is not surprising since algal cells constitute only about 5–15% of the thallus in most cases. It is only when cyanobacteria are present, as in gelatinous lichens, that the proportion together with the gelatinous matrix may rise to 50% or more.

In a lichen the fungal hyphae branch and fuse (anastomose) forming a network of minute, hair-like threads that constitute the bulk of the thallus. Hyphal threads are divided by partitions (septa) that nevertheless permit the transfer of cell substances from one cell to another (Fig. 8). Cells may have varied shapes, largely depending upon the nature of the hyphae. Where septa are wide apart, the cells are elongated; otherwise, they may be globular or ellipsoid. In tightly woven hyphae, they may become almost angular.

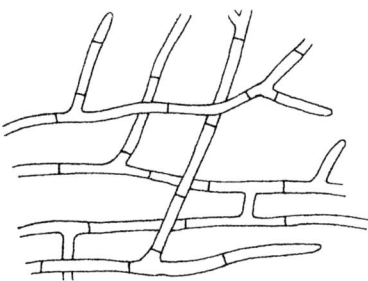

Fig. 8. Hyphal threads showing septa, branching and cross-connection between hyphae.

Most lichen species have an upper layer or 'cortex' of closely compacted hyphae (Fig. 1). Foliose lichens, have a similar lower layer (Fig. 18). The cortex, whether upper or lower, serves as a protective shield, which prevents the photobiont losing too much moisture or being exposed to too much sunlight. It is sufficiently permeable to permit water and gases to enter allowing photosynthesis to take place. It also has a supporting role in maintaining the shape of lobes and podetia. Some of the larger lichens are strengthened by vein-like ridges and the central cartilage-like core of *Usnea* species gives rigidity to the thallus structure (Fig. 20).

The cortical tissue displays various surface features. These include a mat of tiny hairs (tomentum) giving a felt-like covering. This is formed by cortical hyphae thrusting through the surface of the thallus and developing as a very fine down. It probably helps to slow the air flow over the surface of the lichen thereby reducing water loss and may also aid the absorption of water when the lichen is wet, or even facilitate water evaporation if the lichen is oversaturated. Sometimes the surface of the lichen has a powdery film 'pruina' resembling the bloom on grapes. Pruina consists of dead hyphal cells mixed with carbonates or calcium oxalate that accumulate on the surface of the thallus. The function of this layer is not fully understood but it probably discourages insects, etc. from eating it and may be a means by which the lichen excretes excess minerals.

The layer containing the algal cells, intermixed with the fungal hyphae, is situated beneath the outer cortex to obtain the maximum light for photosynthesis. Where photobiont cells, whether algal or cyanobacterial, form this distinct layer within the thallus this is termed 'heteromerous' or layered (Fig. 1). Where the photobiont is distributed among the hyphae throughout the thickness of the thallus, this is termed 'homoiomerous' (Fig. 2). Gelatinous lichens are homoiomerous and have cyanobacteria in which the cells often form distinctive chains. They have neither soredia (see page 28) nor do they produce the 'lichen substances' characteristic of the many species which have green algae. Having no proper cortex and being gelatinous, these lichens may absorb large quantities of water very readily.

Beneath the algal layer in heteromerous lichens is the medulla, composed of loosely interwoven hyphae, which serves as a storage area for both water and the sugars produced by the photobiont. The loosely arranged fungal hyphae allow the diffusion of gases through the lichen.

Lichen colouring

The predominant colours of lichens are grey or whitish grey or green. Brown, yellow and orange are also common and occasionally lichens have a reddish or purplish hue. When wet, lichens with algae may reveal a greenish tinge when their cortical layer becomes semi-translucent allowing photosynthesis to take place. Homoiomerous lichens exhibit a dark blue-grey or brownish-grey colour when wet. There is some correspondence between substrata and thallus colours. Whitish and orange crustose lichens are often found on calcareous rocks; grey or yellow-green colouration is common on siliceous rocks or somewhat acid bark; dark gelatinous lichens are characteristic of microsites with high humidity; yellow or orange pigmentation in the thallus is usually indicative of substrata exposed to high light levels possibly with a nitrogenous content. Such sites are roofs, walls and fences exposed to bird-droppings. In orange foliose lichens, as a general rule, the stronger the light to which a lichen with alga is exposed, the deeper its colouration. This may be noticed when, for example, a lichen thallus spreads from a brightly illuminated horizontal surface to a more shaded vertical one. The orange pigment filters out excessive light (especially UV) which might damage the photobiont. Conversely, in shaded conditions the photobiont requires all the light it can get and in these conditions the orange pigment may be restricted to the fruiting bodies.

Outgrowths from the thallus

Outgrowths from the thallus include spore-producing bodies (apothecia, perithecia, mazaedia and pycnidia) which in some cases are partially or wholly submerged within it and vegetative reproductive bodies (including isidia, soredia, schizidia and lobules). These spore-producing, vegetative reproductive and dispersive bodies are dealt with in the following chapters.

Cephalodia

These are growths that have been likened to galls in plants. They are found in a small number of genera, either on the upper or lower surface of the thallus or in some cases immersed in it. Their distinctive feature is that they contain one or more photobionts that differ from the one which is found in the rest of the thallus. Most commonly these are cyanobacteria of the genera *Nostoc*, *Scytonema* or *Stigonema*, but there are instances of cephalodia containing an alga which develops on a thallus which hosts a cyanobacterium. In some species (e.g. *Solorina spongiosa*) containing cyanobacteria it appears that the fungus is unable to produce fruiting bodies unless a green alga is also present. This green alga may just form a narrow collar around a fruiting body.

External cephalodia take the form either of circular or irregular crinkled patches protruding from the thallus surface or of emergent branched fruticose structures. On the reverse side of the thallus they appear as small swellings. They are thought to result from an intruding photobiont arriving on a thallus and being held by emerging hyphae. More hyphae become involved and a cephalodium forms. Cephalodia cannot be considered parasitic, since they do not seem to harm the lichens in which they develop. Indeed, their relationship is another manifestation of symbiosis, since cyanobacteria, as well as being able to photosynthesize, can fix nitrogen and hence contribute further to the well-being of the host lichen.

Cyphellae and pseudocyphellae

Cyphellae are small rounded or oval hollows or depressions (about 0.5–1 mm wide). True cyphellae are found only on the underside of the species of the genus *Sticta*. They are lined with compacted medullary tissue and are believed to serve as pores aerating the thallus. Pseudocyphellae are similar, but tend to be more elongated with less regular margins. Their medullary hyphae may protrude as tufted outgrowths; alternatively, they may appear as faint white lines or marks on the surface of the thallus where the outer cortex is thin. They are more common than cyphellae, being found in many genera including *Cetraria*, *Cetrelia*, *Parmelia* and *Pseudocyphellaria*.

There are other forms of breathing pores for aerating the medullary layer. In *Parmelia exasperata* they take the form of warts, linked with the medulla, and are composed of hyphae with air spaces between them.

Anchor attachments

Most lichens grow and develop best under a particular set of circumstances and it is therefore important for them that they are securely fastened to the substratum and able to remain in this situation. Some, such as the reindeer 'mosses' (actually lichens), may rely on growing entangled in the surrounding vegetation but most have some means of anchoring themselves. Crustose lichens do this by means of hyphae emerging from the medulla and passing into the immediate surface of the substratum, that is between the cells of bark or timber, or by penetrating a saxicolous surface sometimes to a depth of several millimetres. There are even a very few lichens in the world that are completely unattached to the substratum and blow about in the wind.

Foliose lichens develop 'rhizines' (Fig. 9). These consist of bundles of hyphae which may be branched or simple. The position of the rhizines aids identification as does their shape, especially in the genus *Parmelia*. In such foliose lichens rhizines may emerge from many points on the lower cortex. They are root-like in appearance but, unlike roots, cannot actively draw in water and mineral salts although these may be taken into the thallus by capillary action. Other foliose lichens do not produce rhizines but attach themselves by fungal hyphae over all or part of the substratum covered by the lichen. In umbilicate and most fruticose lichens the fungal hyphae at the base fuse together to form a short, stem-like 'holdfast' which makes a substantial anchor. Holdfasts serve to give strong support to lichens which, because of their protruding thalli on such sites as rocks and trees, are subject to wind pressure.

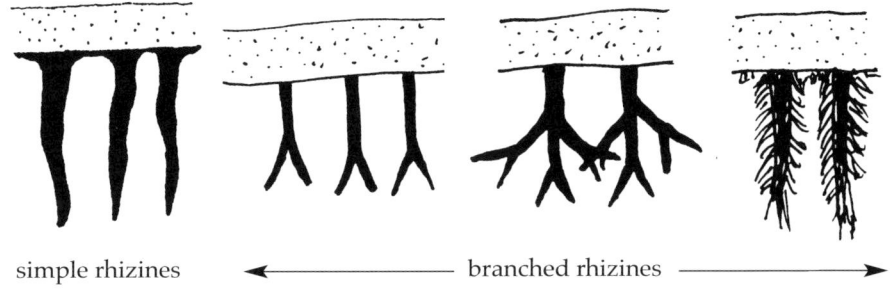

simple rhizines ← ——————— branched rhizines ——————— →

Fig. 9. Anchor attachments

Cilia

These are white, green, black or piebald, thickish, eyelash-like growths, that emerge mainly from the margins of the thallus in foliose lichens. They are composed of adhering hyphal strands which cannot act as vegetative propagules, as they are without a photobiont (Fig. 10).

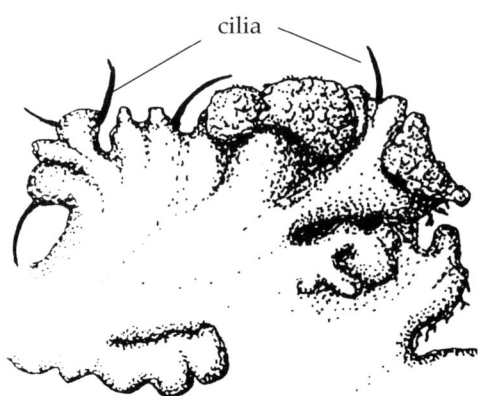

Fig. 10. Cilia on the margin of a foliose lichen

Fibrils

These are short spinulose tendrils, found mainly extending from the stems and apothecia of such genera as *Usnea* (e.g. *U. florida* pl. 4). They differ from cilia in that they contain the photobiont and are capable, therefore, of acting as vegetative propagules (Fig. 11).

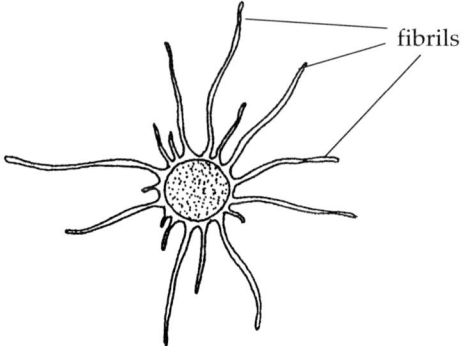

Fig. 11. Aphothecium of *Usnea florida*

Chapter 3
Growth Forms of Lichens

Lichens have a very wide range of forms. Some are leaf-like growths adhering closely or loosely to tree trunks, stone or brick surfaces and are occasionally found on soil and man-made materials such as rubber, glass and lead; some are long, hair-like skeins dangling from branches of trees, or tough, semi-rigid thongs anchored to rock or bark; others are thin or sometimes almost invisible crusts on substrata and may actually be living mainly within the rock or bark. Yet others form cushions or clumps of delicate interwoven branches or stalks, generally on earth or peat.

Leprose lichens

Leprose lichens (Figs. 12 and 13) are those which have not, at least up to the present, been found to have fruiting bodies and so cannot, with certainty, be ascribed to a particular genus. If they are ever found fertile this would allow them to be assigned to their correct genus. They frequently have a complex chemical content suggesting that they are not 'primitive' but have affinities with more highly developed genera. They form a very varied group. Such lichens consist of rather shapeless white, green, grey or yellowish crusts formed of powdery granules (Fig. 13) made of of loose fungal hyphae and algal cells which lack an outer cortex, although some may develop a medulla and weak lobes at the margin. They are often found in damp, well sheltered habitats. In Europe, *Lepraria* (pl. 3) is the most common leprose genus.

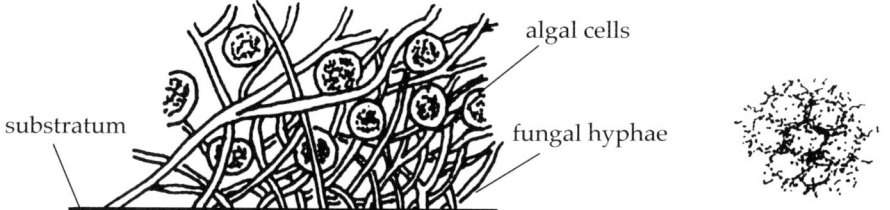

Fig. 12. Section through a leprose lichen x50   Fig. 13. Surface view x15

Crustose lichens

Crustose lichens (Fig. 14), as the name suggests, usually form a crust which adheres closely and totally to its substratum. In some cases the crust is thick and lumpy and may be detached, in part, or may even be submerged below its surface. Often the thallus is discernible only because of the discolouration of the stone or bark. Some crustose lichens have thalli consisting of scattered or loosely grouped granules. They differ from the leprose lichens in having an upper cortex and their algal cells arranged just under this cortex.

In many crustose lichens the thallus presents a patchwork or crazy-paving appearance. The patches or 'areolae' of which it is composed may be relatively large (up to one or more square centimetres) or be very small and raised, thus almost wart-like. The thallus surface may be smooth but sometimes is broken by 'rimose' cracks resulting from thallus surface shrinkage caused by alternate wetting and drying. In some species there is an underlayer of fungal hyphae (the hypothallus) which may be exposed to form a dark rim to the areolae or the extending outer margin of the thallus itself (the prothallus). The thallus is usually firmly attached to the substratum by these fungal hyphae and, in most cases, is difficult to remove without the substratum on which it is growing.

Fig. 14. Section through a crustose lichen

Some crustose lichens have no clearly defined boundaries but others form neat circles. Some species produce well defined lobes around the perimeter (e.g. *Caloplaca* pl. 1). These species are called 'placodioid' (Fig. 15). In other species the fruiting bodies are drawn out into narrow slits so that they look like writing. These are called 'lirellate' (e.g. *Graphis* pl. 2) (Fig. 16). Thalli of corticolous lichens on tree trunks or branches may be oval, they are stretched because the lateral rate of growth of the bark exceeds that of its vertical growth. In crustose lichens growth takes place around the margin, the centre often being reserved for reproduction. This centre sometimes dies and falls out leaving a bare patch which may be colonized by propagules from the original lichen or even by another species.

Fig. 15. Margin of a placodioid lichen

Fig. 16. Lirellate apothecia

Squamulose lichens

Some crustose lichens peel up at their outer edge and form 'squamules' (e.g. *Cladonia* pl. 1 or *Toninia*) (Fig. 17). These may be differentiated from foliose lichens by the lack of a lower cortex so that the loose medulla is exposed often giving an arachnoid (web-like), and in a few cases, a sorediate (powdery) appearance.

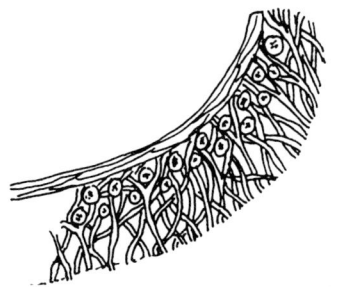

Fig. 17. A squamulose lichen

Foliose lichens

Foliose (leaf-like) lichens (Fig. 18) differ from crustose lichens in having an upper and a lower cortex and therefore may be more easily removed from the substratum. They are attached by root-like 'rhizines' or just by random fungal hyphae. The algal layer is found only under the upper cortex. They usually grow radially, as closely defined rosettes though their outline may be modified by the nature of their substratum and they may form ragged or indeterminate patterns. In almost all cases, they grow mainly at their marginal lobes (e.g. *Parmelia* pl. 3) and the centre may become almost crustose. Exceptions are lichens anchored only at a central point, such as *Lasallia* (pl. 2), where the growth is mainly from the centre. Some foliose lichens are only partially appressed to their substrata and their lobes curl up freely (as in *Lobaria pulmonaria* pl. 3), others are firmly attached, save at their margins, as in most *Parmelia* species.

Fig. 18. A foliose lichen

Fruticose lichens

Fruticose or 'shrubby' lichens differ from foliose species in their bushy form, attached only at the base, but most importantly in the fact that the algal layer is continuous right round the circumference of the the branches of the lichen, just under the cortex. There are a number of foliose species which have the growth form of fruticose species (e.g. *Evernia prunastri*) but if examined closely, it will be found that the alga layer is not continuous over the lower surface. Fruticose lichens may be fine, round, hair-like and loosely attached to rocks and trees (e.g. *Bryoria*) (Fig. 19) and as well as being bushy, fruticose lichens can be flattened and strap-like (e.g. *Ramalina* pl. 4).

Usnea species differ from all other fruticose lichens in Britain in having a tough central core which gives some degree of support to the lichen and which also provides storage for a large proportion of the water contained in the lichen (Fig. 20). *Usnea* species consist of stringy strands which may be short, tufted and bush-like, or reach a length of several feet or even yards. They festoon the branches of trees and rocks and may be so dense as to be almost veil-like.

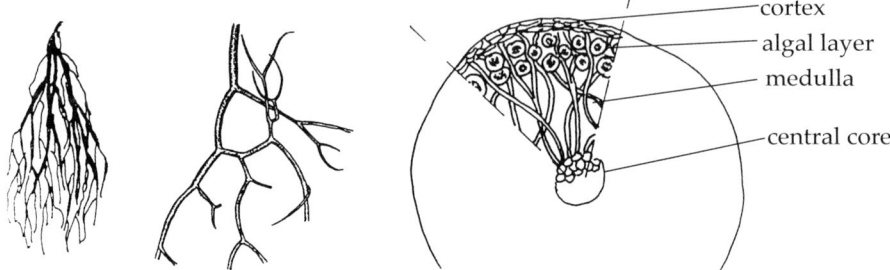

Fig. 19. Two typical *Bryoria* species   Fig. 20. Cross-section through an *Usnea*

Cladoniform or dimorphic lichens

The types of growth forms described in this chapter show our desire to classify every living organism many ways. Unfortunately nature often produces intermediates that will not conform to these man-made groupings. An example of this is found in the genus *Cladonia*. These lichens consist of a primary thallus of minute granules or small squamules, more or less closely appressed to the substratum, and a secondary thallus consisting of fruticose outgrowths. The primary thallus often withers away leaving only the secondary thallus. The latter may form a richly-branched, matted tangle of intertwined branches as in reindeer 'mosses' the *Cladina* group (Fig. 21) or consist of stalk-like outgrowths. These fruticose stems and branches are termed 'podetia' which may be stalk-like or wine-glass-shaped (Fig. 22). The spore-containing bodies are produced on their tips.

20 mm

Fig. 21. A 'Cladina', Cladonia ciliata

Fig. 22. Typical shapes for podetia in *Cladonia*

Filamentous lichens

Filamentous lichens are hair-like and mainly found in damp sheltered sites. They consist of a chain-like alga around which is closely wrapped a thin layer of the hyphae of the fungus (e.g. *Racodium*) (Fig. 23).

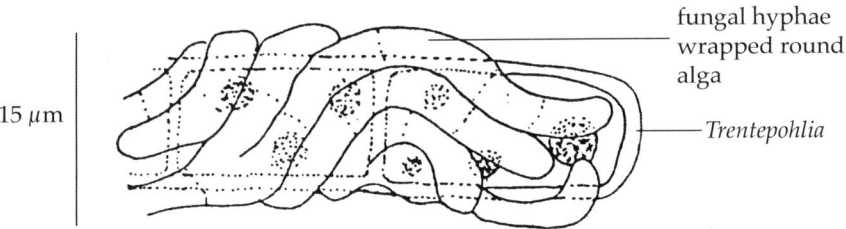

Fig. 23. A filamentous lichen *Cystocoleus ebenus*

Chapter 4
Reproduction and Dispersal in Lichens

Lichens reproduce either by means of spores, or vegetatively. Such a duality is also a characteristic of free-living fungi and many plants, propagation being common either by seeds or by suckering, budding etc. There are three common spore-bearing structures in lichens, apothecia (Fig. 24), perithecia (Fig. 29) and pycnidia (Fig. 32). There is also a variety of types within these groupings such as in the Caliciales (e.g. *Calicium*) where the asci and their supporting tissues quickly break down to a powdery mass (Fig. 30). The rare British members of the Basidiomycota (e.g. *Omphalina*) have cushion-shaped basidia (Fig. 4) on the surface of which the spores are born.

There are two very common vegetative structures, isidia (Fig. 34) and soralia (Fig. 35), and a variety of other propagules. There are also a number of lichen genera (e.g. *Lepraria*, *Thamnolia* and *Racodium*) in which fruiting bodies have never been found. It is assumed that dispersal takes place in these species simply fragmentation of the thallus.

Spores and spore-bearing structures

Virtually all lichens produce their spores in sac-like structures termed 'asci' (Fig. 3). The fruiting bodies containing the asci are known either as apothecia, where they are more or less open, or as perithecia when they are enclosed in a flask-like body. Many non-lichenized fungi have fruiting bodies which persist for just a few days whilst in lichens individual apothecia and perithecia often continue to produce spores for up to 30 years in some cases .

Apothecia

Most apothecia, though by no means all, take the form of an exposed disc 0.1 mm to 1 cm or more in diameter. This disc may, in some cases, be supported on a wine-glass-shaped structure (a scyphus Fig. 22) or it may be sitting on the surface of the thallus (sessile), be level with it or be sunken below (immersed). In other cases, the disc or globose apothecium may be borne on a stalk (a podetium, Fig. 22). The asci are supported and surrounded by sterile hairs (paraphyses). (See below under Hamathecia and Asci.) The shape and form of these paraphyses is often important in identification. They may have swollen coloured tips and it is this pigment, when viewed from above, that gives the disc its colour.

Fig. 24 shows the main features of apothecia and illustrates the differences between lecanorine and lecideine types. These two main grouping were considered of great value in the classification of lichens but modern techniques have lessened their importance. However, in studying lichens, it is still very useful to understand the difference between them.

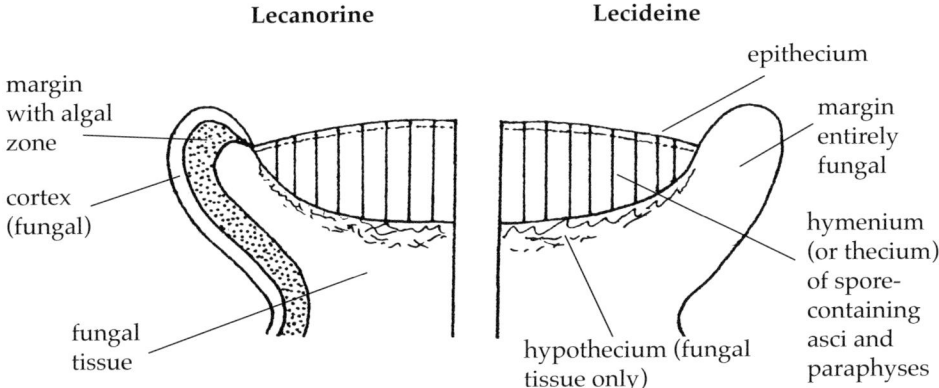

Lecanorine

Lecideine

epithecium

margin
with algal
zone

cortex
(fungal)

fungal
tissue

margin
entirely
fungal

hymenium
(or thecium)
of spore-
containing
asci and
paraphyses

hypothecium (fungal
tissue only)

Fig. 24. Lecanorine and lecideine apothecium structure

In lecanorine apothecia (e.g. *Ochrolechia parella* pl. 3), the algal layer continues into the rim of the disc (thalline margin). Thus the margin to the disc is the same colour as the remainder of the thallus. It should, however, be noted that this margin is often eliminated as the central disc develops.

In lecideine apothecia (e.g. *Rhizocarpon geographicum* pl. 4) there are no algae in the rim of the disc (lecideine or proper margin). The margin of the disc is, therefore, normally a different colour from the thallus .

In some genera, apothecia take the form of thin narrow ridges or splits in the thallus, with the 'disc' taking the form of a narrow slit. These are termed 'lirellae' (e.g. *Graphis* Fig. 25, pl. 2). Apothecia may develop on the lobe tips of some foliose lichens, on the upper surface in *Peltigera* (pl. 2) and on the lower surface in *Nephroma*. In some other genera (e.g. *Baeomyces*) the apothecia resemble miniature mushrooms (Fig. 26) with the fruiting body located on the top of a short stalk (a stipe). This stalk is normally colourless as it does not contain the algal partner. The disc may be smooth, matt, pruinose or not. It may also be contorted and wavy or covered by wavy folded ridges (e.g. *Umbilicaria*) termed 'gyrose' (Fig. 27).

Fig. 25. *Graphis* species   Fig. 26. *Baromyces rufus*   Fig. 27. A gyrose apothicium

Some genera of lichens (e.g. *Arthonia*) do not develop a firm margin and the hymenial tissues emerge directly from the thallus, tending to give the fruiting body a somewhat irregular shape (Fig. 28).

Fig. 28. An arthonioid apothecium

Perithecia

The fundamental difference between apothecia and perithecia is that in the latter the hymenium is wholly or in large part enclosed in a globular or flask-shaped structure, frequently composed of hard carbonaceous tissue, usually black. On the upper surface of the perithecium there is a tiny opening or 'ostiole', through which the ripe spores escape (Fig. 29). Superficially, the difference in appearance between apothecium and perithecium is not always clear. For example, in the genera *Diploschistes* and *Thelotrema* and in some species of *Pertusaria* the margin of the apothecium grows to cover much of the disc, which then presents the outward appearance of a perithecium.

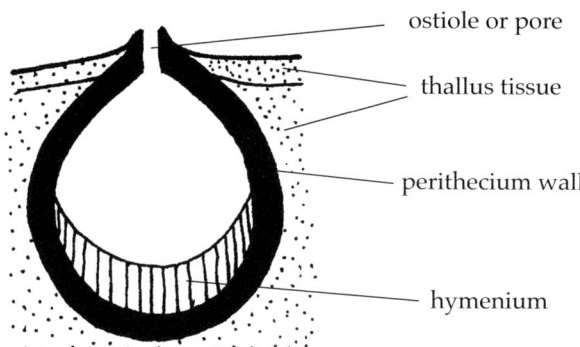

Fig. 29. A diagramatic section through a perithecum

Perithecia may be 'endolithic' or 'endophloeodal', meaning total immersion in the substratum of rock or bark respectively, except for their tiny ostioles. In others, immersion is less complete and the carbonaceous envelope or involucrellum protrudes beyond the thallus surface. Because of their tough outer casing perithecial lichens are classified as pyrenocarps, i.e. as bearing hard nut-like fruits. Sometimes, where the perithecia have eroded tiny pits in the rock, the dead remains are caught inside whilst in others the involucrellum is merely a 'trap-door' which falls off at death leaving an empty pit.

Finally, there is a small group of lichens (e.g. *Calicium* and *Coniocybe*), which carry their spores in asci which disintegrate before the spores are fully ripe. The result is that spores and paraphyses together make an undifferentiated, dry powdery mass known as a 'mazaedium' (Fig. 30). This minute structure takes the form of a tiny knob carried, in many genera, on the top of a thin stalk.

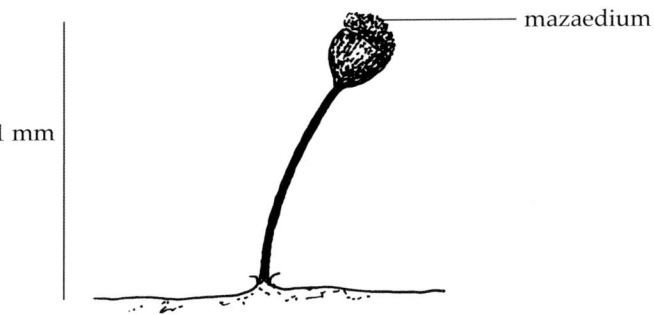

Fig. 30. Stalked mazaedium

Hamathecia and Asci
The hymenium (Fig. 24) in both apothecia and perithecia is composed of the spore-bearing asci and the tissue formed of fungal hyphae that lies between them. This tissue is termed the 'hamathecium'. The sterile hyphae (paraphyses) originate from the base of the hymenium. They may be simple, branched, or they may anastomose to form a loose network. Alternatively, they may be glued together or break down early in the development of the fruiting body.

In perithecia these hyphal elements develop in several different ways. They may grow downwards to the base 'pseudo-paraphyses' or only for a short distance from the sides of the perithecia (periphysoids) or they may originate in the upper parts of the perithecium and grow upwards towards the ostiole (periphyses).

In most lichens the ascus has a more or less club-like form, with the thicker end towards the surface of the apothecial disc or the perithecial ostiole. Sometimes it is cylindrical with the spores arranged in a single row down its length, or may bulge to give an almost lamp-bulb outline. The shape of the ascus is important in classification and emphasis has been placed, recently, on the structures contained within it and on its surface. These may be seen when they turn blue after treatment with potassium hydroxide followed by iodine.

Ascospores
Spores contain one or more nuclei and other elements enveloped in a strong cell-wall made of chitin (a substance also found in the very tough outer skeleton of insects). Ascospores are initially unicellular and colourless and many remain so. Others develop dividing walls (septa) and some darken becoming brown or greenish.

There are great variations in spore size, from those of the genus *Acarospora* (4 μm x 2 μm), to those of the genus *Pertusaria* (250 μm x 80 μm). Some are even larger in certain foreign species. Spore size varies within genera and is not completely uniform within species. Spore shape varies also, some being ovoid and others extremely long and slender. There may be a thick gelatinous covering (perispore) or the outer wall may be rough or with scaly protrusions (ornamented). Spores are generally classified as simple, septate, muriform or polarilocular. Within these main types there is great variation even within the same genus as regards number of septa and muriform divisions, as well as in size and colour. Degrees of curvature may vary also. Especially great are differences in the form of the cells and the connecting canals of polarilocular spores (Fig. 31).

Simple spores (unicellular)

  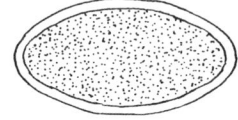

Thin-walled e.g. *Lecidia* species   Thick-walled e.g. *Pertusaria* species

Septate spores (with divisions)

Single division e.g. *Buellia* species   Multiple divisions e.g. *Graphis scripta*

Muriform spores (with both transverse and longitudinal divisions)

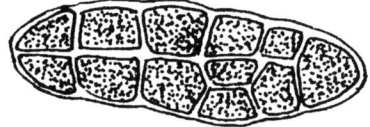

e.g. *Rhizocarpon geographicum*

Polarilocular spores (two celled with a linking channel)

  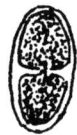

e.g. *Caloplaca* and *Xanthoria* species

Fig. 31. Examples of ascospore types

There are usually eight spores per ascus, although one or more may be undersized or fail to develop at all. Within the ascus spores may overlap, be serially arranged or, when very numerous, form a tightly packed mass. When spores are tiny, as in the genus *Acarospora*, they may be counted in their hundreds but when the spores are very large, as in the genus *Pertusaria*, there may be only one or two.

The number, size and wall thickness of spores illustrates the different strategies employed by lichens to ensure that some spores germinate successfully to produce a new thallus. Large numbers of very small spores mean that there is a good chance of one or more landing on a suitable habitat. However, the thin walls and the small supply of food in the spore allows only a short period of viability. Alternatively, large, thick-walled spores, of which there may be very few in each ascus, are able to survive for long periods while awaiting the arrival of a suitable alga.

Ascospore structure in lichens is similar to that of non-lichenized fungi but differs from them in one striking respect. They are not fully reproductive agencies of the lichen since they do not incorporate the essential photobiont component of the lichen symbiosis. They can germinate and may survive for a short time producing fungal hyphae but without an appropriate photobiont they cannot develop to form a complete lichen.

How spores, once released from their parent plant, encounter and come into association with compatible photobiont cells is still a major question in lichenology. Answers are mainly conjectural. Chance encounters may result in fusions, more particularly where a common alga (e.g. *Trentepohlia*) or a common cyanobacterium (e.g. *Nostoc*) is involved. This seems less likely when, for example, the photobiont is *Trebouxia* which is rarely found, except in lichenized fungi. It is probable that new lichens depend on algal cells picked up from established lichens in the vicinity, either from soredia or isidia or from dispersed thallus fragments. This view is supported by the tendency for a particular species to disappear once its population density declines in a locality, thus reducing the probability of an appropriate fusion of alga and fungal mycelia to form a new lichen. It is certain that many lichens are parasitic on other lichens and that a germinating spore may take over the alga from the host. In other lichens, spores are able to lie dormant and await encounter with the needed alga and in yet others the spores may produce mycelia that can live saprophytically on other material for a time surviving as an ill-defined protolichen. There is much reasonable conjecture, but also growing substantiating evidence which is based on research and experimentation.

Spore discharge results from the ascus tips splitting apart on maturity. The exact mechanism for this varies depending on the type of ascus involved. It occurs as the ascus becomes distended as its spore content expands. Other factors that influence the process are increases in humidity and the occurrence of the right temperature levels. Increasing humidity causes the hymenial contents to swell, increasing the pressure in the ripe ascus. This is a continuing process as new asci are developing constantly. Experiments show that discharge appears to take place more freely in darkness following exposure to light.

Dispersal experiments have shown that spores in a windless closed system may be ejected upwards from 3 mm to 45 mm which would be sufficient to bring them into contact with moving air currents. Wind then becomes the agency for longer-distance transport. Dispersal may also be by insects, mites or larger animals and by rainwater splashing over lichen surfaces. The slugs and snails which graze on lichens may play a significant role.

It is very difficult to observe germination in the field but it has been shown that although lichens do not exhibit the seasonal variations seen in the higher plants, spore discharge and dispersal take place most frequently in the autumn and spring in temperate zones.

Conidiomata

The term 'conidiomata' is used to describe the bodies in which conidiospores are produced. In almost all lichens these are in the form of pycnidia (Fig. 32). These resemble perithecia superficially and can at first sight be confused with them or, indeed, with the fruits of certain lichenicolous fungi that are parasitic on lichens. Like perithecia, they are usually minutely globular and have a surrounding wall with an ostiole through which their conidiospores are expelled. Pycnidia appear as tiny black dots on the surface of foliose species (e.g. *Hypogymnia physodes*) and of crustose species. In some species such as *Cliostomum griffithii* they are almost always present, whilst apothecia are rather rare. They may also be spherical to peg-like and protrude above the surface of the lichen (e.g. *Lecanactis abietina*) or develop on the tips of branches or podetia of fruticose species (e.g. *Cladonia furcata*). Internal pycnidia are generally round, but on occasions they may be irregularly formed and divided into small chambers (loculi).

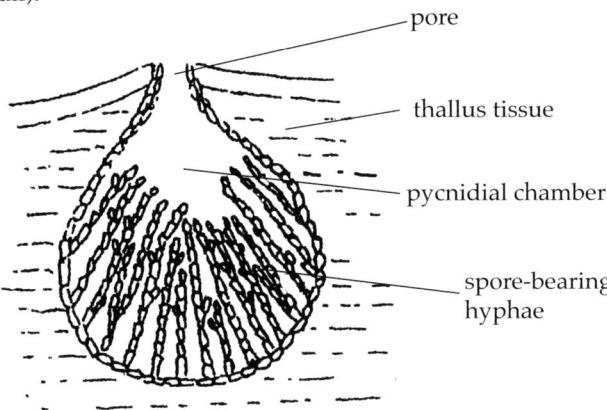

Fig. 32. Pycnidium in vertical section

Conidiospores are not formed in an ascus but emerge from the sides or ends of special hyphal filaments called 'conidiophores' which develop in the pycnidium (Fig. 33). They are generally very small, globose, elliptical, thread-like or sickle-shaped bodies, which are always colourless and are produced in great numbers.

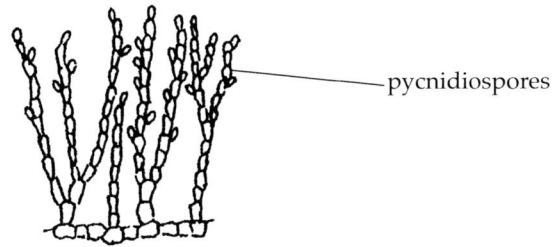

Fig. 33. Detail of spore-bearing hyphae

Two types of conidia are often produced by lichens. The term 'macroconidia' is used for the larger spores which may often have the power to germinate in the same manner as an ascospore. The smaller spores are called 'microconidia' occurring frequently in the same thallus and even in the same pycnidium as macroconidia. There are some species in the genus *Micarea* that even produce three types of conidiospore.

It should be stated here, however, that curiously little has been published about conidia, although they occur in some 8,000 species of lichen. Full use is just beginning to be made of new technology to explore their forms and the possible roles they may play in lichen reproduction.

Vegetative structures and propagules

Vegetative reproduction and dispersal is of the greatest importance to many lichens, since it is the only known means by which many can achieve propagation. It has the advantage over reproduction and dispersal of fungal ascospores in that it contains both partners of the symbiotic relationship. It is, however, asexual and as such does not allow for a new genetic mix to be produced. It may take place by the dispersal of fragments broken off from any part of the thallus in which both fungal and algal components are present, but there are also a number of specialized outgrowths which aid dispersal, the commonest of which are isidia and soralia. Such structures are not found in non-lichenized fungi.

Isidia
These take a number of forms. Some are tiny cylindrical outgrowths, others are globular, and yet others are tiny marginal lobes or coral-like branchings. They are found in some 20–30% of foliose and fruticose lichens but are somewhat less common in crustose lichens.

The most important characteristic of isidia is that they are outgrowths with the outer cortex and algal layer continuous from the thallus, over the isidium and back to the thallus (Fig. 34). Differentiation of isidia from other related thallus outgrowths is of no great significance. Papillae and lobules have a similar composition and also serve as vegetative propagules as do the thread-like fibrillae of the genus *Usnea* (Fig. 11). It has been suggested that isidia may also serve to extend the total thallus surface and hence its capacity for gaseous exchange and photosynthesis.

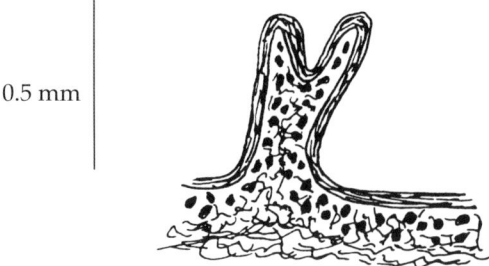

0.5 mm

Fig. 34. Section through isidium

Soredia

A soredium consists of a small number of algal cells enmeshed in fungal hyphae but it differs from an isidium in having no cortical layer and arises from the breakdown of the cortex. Most usually, soredia are found in clusters in a body called a 'soralium', but this is not always the case and they may be scattered over the thallus or on the podetia of some species. They form as fungal and algal cells that come together within the thallus and arise from breaks and often at thin points in the upper cortex. They have the chemistry of the medulla and its distinctive reactions to chemical tests. As pressures develop because of their growth they thrust through the cortical tissue forming soralia. Soralia may also develop when isidia break down, as is the case in some of the brown species of the genus *Parmelia*. The wart-like 'papillae' found on the stems of some *Usnea* species may break down to release soredia.

The position and type of soralia are very helpful in identification. They take a number of forms, as shown in Fig. 35.

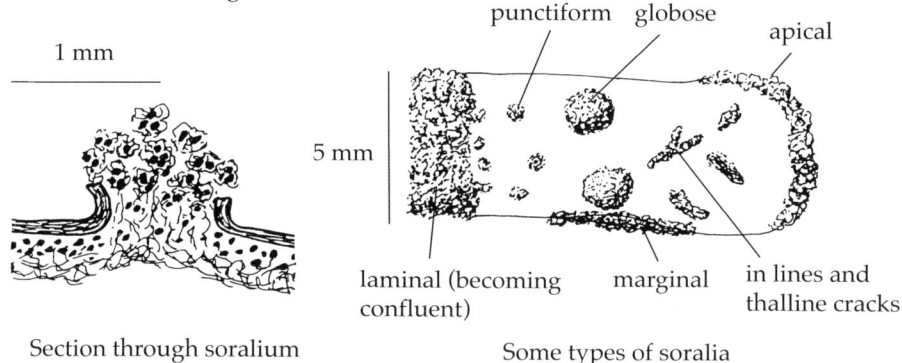

1 mm

5 mm

punctiform   globose

apical

laminal (becoming confluent)   marginal   in lines and thalline cracks

Section through soralium   Some types of soralia

Fig. 35. Soralia

Isidia are found, in general, on lichens favouring sunlight and dry conditions. Soralia, on the other hand, are also frequently found on lichens growing in damp and more shady places. Soredia are dispersed by air currents, wind and by rain droplets but birds and insects also have a part to play. Isidia are heavier and their range of dispersal is consequently much less.

Lobules

These tiny lobes develop on the upper surface or margins of the lobes proper of foliose lichens. Like many isidia they may be constricted at their base and this facilitates their dispersal. Unlike isidia they are flattened. Lobules are not always easily distinguished from isidia, particularly in the early stages of their development, nor from some forms of papillae or fibrillae. They are very similar to the scale-like 'phyllidia' found in species such as *Peltigera praetextata*. Phyllidia are always flattened, dorsiventral and constricted at the base and are, therefore, much more likely to break off than lobules.

Other terms used for vegetative growths include: 'schizidia' (e.g. *Fulgensia bracteata*), fragments of the thallus surface, with algal cells attached, which flake off as scales; 'phyllocladia' (e.g. *Stereocaulon*), the numerous minute leaf-like protrusions found on the podetia of some species of fruticose lichens; 'blastidia' (e.g. *Leacnia erysibe*), small packets containing both partners which bud off the thallus; 'hormocysts', found in the gelatinous genus *Lempholemma* and formed from chains of algal and fungal cells that grow in bodies called 'hormocystangia' before breaking off and being dispersed; 'goniocysts' (e.g. *Vezdaea aestivalis*), minute granular, corticate packets of fungal hyphae and algal cells.

Sexual and asexual modes of reproduction

The sexual process in lichens does not involve the lichen as a whole, since the photobiont, whether alga or cyanobacterium, normally multiplies only by asexual means. In a cyanobacterium this merely involves simple cell-division and in an alga, a specialized cell known as an aplanospore, which does not result from sexual fusion, is often produced. The spores released are, therefore, only from the fungus but in a few genera (e.g. *Endocarpon*) small algal cells of the same species as the parent thallus are found amongst the asci, and it is quite possible that some of these are dispersed with the fungal spores, so facilitating the formation of a new lichen thallus.

Within the fungal mycelium of the developing thallus some hyphae become differentiated from the rest and form ascogonial tissue (i.e. female tissue from which ascocarps later develop). From this ascogonial tissue emerge hair-like filaments (trichogynes) which may 'catch' and transfer a male conidiospore (a spermatia) from a pycnidium to the ascogone of the same lichen. How the two partners to this presumed sexual equation evolve is not known. It has not been shown that they derive other than from the same thallus and there is no indication of some thalli being more markedly male or female than others. Nor is it known how, if at all, sexual fusion takes place in lichens that do not produce pycnidiospores. It is probable that two fungal hyphae develop and then exchange DNA and from this exchange the ascomata develop. The ascocarps, whether apothecia or perithecia, then develop in the ascogonial tissue.

It is assumed, by analogy with non-lichenized fungi, that sexual fusion, in the form of the interchange of cell contents resulting in a doubling of chromosome content, does take place. But it has been pointed out by a well-known American lichenologist, Mason Hale, that the "actual modes of reproduction . . . remain a complex little-understood process, and no other aspect of lichenology is so poorly known and intractable to experimental study."

An important element in this problem results from uncertainty over the role of pycnidia and of the microconidia and macroconidia which they produce. This has been, and still is, a matter of conjecture. Lichenologists in the eighteenth and nineteenth centuries thought that pycnidia solely furnished the male element in lichen reproduction. Hence, they were equated with spermagonia and their spores with spermatia. Recent work has shown that they may have a dual role, serving a sexual function as spermatia or acting as asexual spores.

Little is known about the processes of discharge, dispersal and germination of lichen conidiospores. It has been suggested that raindrops in some cases may transfer extruded conidiospores to protruding trichogynes and thus set in train a sexual form of reproduction.

It is clear, however, that in some lichens there must be both asexual and vegetative modes of reproduction, that is, they have spore-bearing apothecia or perithecia together with isidia or soredia. But these are the exceptions: the majority rely on one mode or the other, at least at any one time. This qualification is necessary as the mode may vary according to the age of the lichen. For example, younger thalli of *Peltigera didactyla* carry only soredia but older thalli have apothecia. On the whole most lichens with wide distributions, such as *Parmelia saxatilis* and *Evernia prunastri*, have vegetative propagules as the primary means of reproduction and dispersal, which has the advantage of combining fungal and algal elements, while spore-dependent species must rely on their spores coming into contact with an appropriate photobiont.

There are some lichens which form 'species pairs' (e.g. *Lecanora intricata* and *L. soralifera*), the one with spore-producing structures, regarded as the 'primary species', the other with vegetative propagules, regarded as the 'secondary species'. In other respects, lichens of both types in these 'species pairs' are identical. It is possible that these two species split apart from a common ancestor in the past but it appears that the primary species is usually fertile and the non-fertile species is derived from it. This asexual species has the advantage of disseminating both partners at the same time. It is not possible to judge how far such lichens benefit in the course of time from the survival value of vegetative reproduction on the one hand, and from the adaptive potential of sexual reproduction on the other.

Hybridization

Lichen species show normal evolution from sexual reproduction. Recent research has shown that variation is also found among thalli arising from vegetative propagules. Tree-bark surfaces have been inoculated with soredia and the resulting growths were studied at regular intervals with a scanning electron microscope. It was found that soredia from different individuals of the same species fused to form a single hyphal net and, more strikingly, that young fragments of the thalli of *Physcia adscendens* (pl. 3) and *Physcia tenella* can merge together to form a lichen combining the characteristics of both species. It is probable that these vegetative or 'mechanical' hybrids are more common than had been previously thought. There are also temporary hybrids such as *Xanthoria parietina* (pl. 4) taking over the alga of a species of *Physcia*.

Chapter 5
Lichen Physiology

Nutrition

Because lichens have a photobiont component, they are able to carry out the process of photosynthesis. Thus, like plants, their nutrition depends on some simple chemicals although in varying degrees. The sources of these, together with the part played by the mycobiont and photobiont partners, are described below.

Carbon is obtained as carbon dioxide from the atmosphere. By means of the chlorophyll present in the photobiont the energy of the sun is used to synthesize carbohydrate from carbon dioxide and water. The fungal component cannot do this as it requires synthesized carbohydrate. It obtains this from the photobiont in the form of glucose from cyanobacteria and mainly ribitol from algae. These are converted to mannitol by the fungus for storage in the medulla.

Nitrogen, which is essential for the synthesis of protein, is obtained mainly by the fungus from bird droppings, from tree bark, or from decaying organic matter in the substratum. Since rainwater or water vapour may be absorbed over all parts of the thallus, nitrates and soluble organic nitrogen-containing compounds can be absorbed by the fungal hyphae and by the photobiont cells. Organic nitrogen in quantity appears to be essential for some lichens, such as *Parmelia conspersa* (Fig. 36), which flourish on bird-frequented rocks but, in excess, it can hinder or prevent the development of others. If the photobiont is a cyanobacterium it can fix atmospheric nitrogen directly. Where this is the case such lichens have a higher nitrogen content than those which have an algal photobiont.

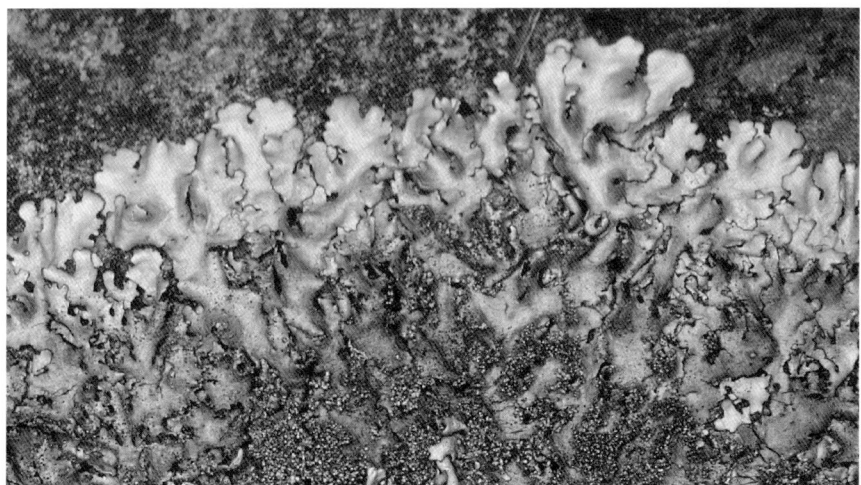

Fig. 36. *Parmelia conspersa*

All other essential elements are absorbed from the environment in the ways described above and are taken up into the cells of both partners of the symbiosis. It has been suggested that very slow-growing lichens can obtain almost enough minerals from those in solution in rainwater. Minerals can also be obtained from the substratum by capillary action and from its breakdown caused by the acid substances produced by the lichen. The wetting and drying of a crustose thallus often causes stresses in the substratum and tiny grains of rock may thereby become incorporated into the thallus. Excessive wetting by rain may leach out a large proportion of the minerals held in the thallus.

Lichens and water

Lichens cannot control their water content, unlike vascular plants which can preserve an appropriate balance despite changes in their environment. It is the level of humidity in the atmosphere and in their substrata that is the determining factor since, just like sponges or blotting paper, they absorb or lose water mechanically and can neither accumulate it nor resist its invasion. This does not mean there are no variations in water absorption and loss in different lichen genera and species. Gelatinous lichens, for example, absorb and lose water at a higher rate than lichens containing alga. The thallus of a foliose lichen can conserve moisture, both by reducing the evaporation from the substratum under their protecting thalli, as well as by storing it internally, especially in the medulla. Fruticose lichens, such as *Cladonia* species are similarly served by their matted branches which cut down the flow of desiccating air over the surface. *Usnea* species (Fig. 20) may conserve up to 30% of their water in their central axis. Crustose lichens are protected from potential water loss through their undersides by their firm attachment to their substrata and, like most other homoiomerous lichens, by the cortex of compacted hyphae. In addition, the presence or absence of rhizines, isidia, tomentum, pruina and other morphological features all have an influence.

Water is taken in through any part of the thallus surface but capillary action may also be important. In the genus *Umbilicaria* and in other foliose lichens, rhizinae can draw in water by this method, whilst *Toninia* species have an unusually long and tough rhizinal 'root' which penetrates the substratum up to a depth of several centimetres coming into contact with moisture well below the immediate surface of the substratum.

Modes of water absorption and loss differ. In crustose lichens water from the substratum may enter the medullary hyphae immediately since these are in direct contact with the substratum, whilst the cracks between areolae are significant for the intake of water from the upper surface. In foliose lichens, where there is a lower cortex between the medulla and the substratum, water is mainly taken in through the upper surface of the thallus from airborne sources. In gelatinous lichens the absorptive capacity is very considerable and the thallus soaks up water rapidly.

When saturated, most foliose and fruticose lichens may hold an additional 100–300% of their dry weight as water, and this is also seemingly true of crustose lichens. Gelatinous lichens can absorb from 800–3,500% of their dry weight. Water content can soon be lost, although not as rapidly as it be acquired. Often no more than several minutes are needed for saturation, or near saturation, to be achieved and for equilibrium with the surrounding atmosphere to be reached.

Another factor enabling lichens to survive in hot, dry conditions is that many do not require surface water or rain. They are able to absorb moisture from the atmosphere and indeed are well adapted to do so. Thus, those which grow in tropical deserts can benefit from night-time condensation, allowing photosynthesis to take place in the early morning, but can rid themselves of their water content before they are heated by the sun to the point when it would destroy their tissues. Atmospheric moisture in the form of mist is favourable to many lichens, including those that thrive on mist-bound coasts and hills in temperate regions like the Lake District and the Western Isles, as well as in mist belts at high altitudes, as on Tenerife and in the Seychelles. Some lichens are virtually impervious to liquid water. It simply runs off their surface. These include the genus *Lepraria* (pl. 3), which is found in shady, damp habitats under roof overhangs or on sheltered woodland trees, and *Bryoria* species which cannot take in water through their strands even if these are immersed. Such lichens depend entirely on atmospheric moisture or damp substratum.

Lichens can withstand substantial periods of drought. It has been shown that *Rhizocarpon geographicum* (pl. 4) can survive for nine months or even more without moisture. Drought is, in general, damaging and the drier atmosphere of large cities may be a contributory factor in their becoming 'lichen deserts'. Conversely, if they become over-saturated, gaseous exchange is difficult and photosynthesis much reduced. The optimum level is 40–70% in most lichens.

Lichens and temperature

Lichens can withstand wide ranges of temperature and hence adjust to seasonal changes to a much greater extent than the higher plants. They can survive on the sun-drenched rocks of the Negev Desert and in conditions of arctic severity. Their slow growth rate and their adaptability are in large measure due to their being able to dry out, yet maintain a low level of respiration.

Investigations have shown that lichens can maintain half of their normal level of respiration at temperatures as high as, for example, 70°C in *Alectoria sarmentosa* and 101°C in *Cladonia pyxidata* (pl. 1). This is only true if they are air-dry. If thalli are moist, their respiration virtually ceases at temperatures approaching 45°C. Resistance to cold is still more remarkable. Some species can revive after several hours of exposure at minus 190°C. One species of lichen survives in the Antarctic by living beneath the translucent surface layer of rock where it still receives sufficient light for its alga to photosynthesize. Even when fully saturated and thus more exposed to the disintegrating effects of freezing, there have been cases of lichens surviving for two days at minus 78.5°C. Crustose and foliose species, adhering to bare rock or bare ground in hot sunshine or deep frost, may be far warmer or colder than the surrounding atmosphere.

Lichens and light

Lichens which rely on algae or cyanobacteria must have light for the process of photosynthesis by their photobiont. They cannot, like some fungi, thrive in darkness and are not found, for example, deep in caves. Light, indeed, is a major factor in determining the viability of lichens, their form and distribution. Their requirements are even greater than those of plants since they have, weight for weight, between four and ten times less chlorophyll. Some research shows that the algal layer in the thallus of a lichen may become thicker in lower light situations.

Some lichens are found only in full sunlight (e.g. *Parmelia conspersa*). Other lichens favour situations in which light is diffuse such as shaded or partially shaded tree trunks or rock surfaces, the north walls of buildings or under the overhangs of rocks (e.g. *Trentepohlia* containing lichens such as *Dirina* and finely filamentous lichens such as *Cystocoleus*). Sunlight appears to compensate for low temperatures. Lichens which flourish in bright light are a feature of cold, exposed northern and high altitude regions.

Light intensity affects the colour of many lichens. *Xanthoria parietina* (pl. 4) is bright yellow in sunny situations (e.g. on exposed roofs), but greenish-grey in shade (e.g. on walls under trees) whilst *Rhizocarpon geographicum* (pl. 4) is yellowish green when exposed to strong mountain sunlight, but greenish where the light is weaker. These variations would seem to be due mainly to an increase in pigmentation but also to a thickening of cortical tissue, both of which serve to protect the photobiont from over-exposure and from the desiccation resulting from it. It is interesting to note that lichens have a capacity for regulating the effects of an environmental factor such as light that they do not possess in the case of water. Water content, however, determines the permeability of the cortex to light. As the water content increases the upper cortex swells becoming translucent, thus allowing more light to reach the photobiont.

Where lichens survive in conditions of light deprivation, they not only lose colour, but they often become sterile. Those lichens that appear to be always sterile often flourish in situations of low illumination.

Lichen substances

Lichens are unusual in that they produce chemical substances seldom found elsewhere. Beginning some 100 years ago, there is an elaborate and expanding interest in lichen chemistry, which has its own distinctive technology. An advanced knowledge of biochemistry is required for a full understanding of this area and only indications of its content and significance can be given here.

A distinction must be made between substances produced within the cells of lichen hyphae, termed intracellular, and those produced outside such cells, termed extracellular. Intracellular substances, including both proteins and amino acids, and carbohydrates and vitamins of various kinds, are found in all living organisms but it is the extracellular substances excreted by the fungal hyphae that have come to be termed 'lichen substances'. They have been detected almost solely in lichens and are confined, except for a small number found in fungi, to lichens with algae. They number at least 350, but the chemical structure of all of them has not yet been fully determined. They include acids (such as the common usnic acid), esters, depsides and depsidones and are often found as crystalline deposits on hyphal surfaces. They are often the determiner of lichen colouration. The orange of the Teloschistaceae (e.g. *Xanthoria* and *Caloplaca* pls. 1 and 4) comes from the presence of an anthraquinone, called physcion, in the cortex. Calcium oxalate often produces a whitish glittering pruina on the thallus surface. Very often a different substance is produced in the medulla to that in the cortex.

Although lichen chemistry has not replaced the long-established reliance on anatomical and morphological features for identification it has become an important element in lichen taxonomy and is often useful to separate similar-looking species. It has been suggested that chemical diversity does not always reflect genetic differences but may be due to environmental factors. Thus chemical variants are not considered adequate in themselves to justify the creation of separate species. It is accepted that the same lichen may have two or more 'chemical races', that is, variations in the chemical substances it produces. Furthermore, a taxonomy relying heavily on chemical testing has disadvantages. As Hawksworth writes, "Lichen chemotaxonomists are not the only botanists who use lichen names and so lichenologists must be wary of devising taxonomic systems which cannot readily be used by others. A situation in which, in order to name a specimen to species rank, an ecologist has to examine much of his material both by thin-layer chromatography and study it under the scanning electron microscope is not perhaps in the best long-term interests of lichenology, but may be being approached in some cases." (Hawksworth: *Problems & Progress*, 1976, p. 156)

Extensive studies have been made in identifying lichen substances, but their functions have not yet been fully established. It is conjectured, however, that where they are responsible for pigmentation, they serve to offer protection against excessive light. It is also suggested that they may protect lichens from bacterial and fungal invasion and give some defence against browsing slugs, snails and insects. Lichens do seem more resistant to various infections and mites than other plants even though damage caused by such creatures is often very obvious. Medical uses of lichens and the qualities that have promoted their use in dyeing and perfume manufacture are usually related to the lichen substances which they produce. They also appear to have a role in 'chemical warfare' for occupation of the substratum. Where they are produced at the margins of the thallus they may prevent the encroachment of neighbouring lichens. These may even be of the same species but slightly different genetically. They also inhibit the growth and germination of some higher plants. In North America dense swards of *Cladonia* (pl. 1) beneath trees appear to result in a thinner canopy.

Growth, decay and death

Growth in lichens is largely on the margins of the thallus except in umbilicate species (e.g. *Umbilicaria*) where it is from the centre. In fruticose species it is on the tips of the stems, branches or podetia and in filamentous species as a lengthening of the hair-like strands of which they are composed. There is little central growth save that in some, the thallus may thicken. This leaves the centre better able to produce apothecia, perithecia or pycnidia and vegetative propagules such as isidia and soralia. In a number of cases reproductive structures, such as the apothecia of *Peltigera* species, are marginal.

In foliose lichens, as well as the growth of marginal lobes, additional lobes may be formed in the centre of the thallus. In crustose lichens, the origins of the lichen may be in the form of small separate islands (areolae) which coalesce to form the thallus or a continuous thallus may grow from one central point and become randomly cracked due to the wetting and drying of the surface (rimose cracking). Some lichens have zoned margins, for example species of *Buellia* and *Ochrolechia* (pl. 3) representing periods of rapid and slow development. Growth also occurs, sometimes, as a renewal of the thallus centre which may have fallen out, allowing regrowth by the same or another lichen (e.g. *Diploicia canescens* pl. 2).

The greater part of the thallus comprises the fungal component. As it grows there must be a corresponding increase in algal cells to maintain the balance. It is not clear how this occurs but it has been seen that, near an actively growing tip, the algal cells are small and dividing rapidly, whilst further into the thallus the algal cells divide much more slowly and are larger.

In general, lichens are extremely slow-growing plants, especially in harsh conditions of deserts and Antarctic wastes where the growing period may be very short. Here the growth rate is often less than 1 mm a year. Rapidly growing species include *Usnea articulata* in the British Isles and, in the ideal growing conditions in the coastal mist region of California, the 'lace lichen' *Ramalina menziesii* (Fig. 38) are measured in centimetres per year.

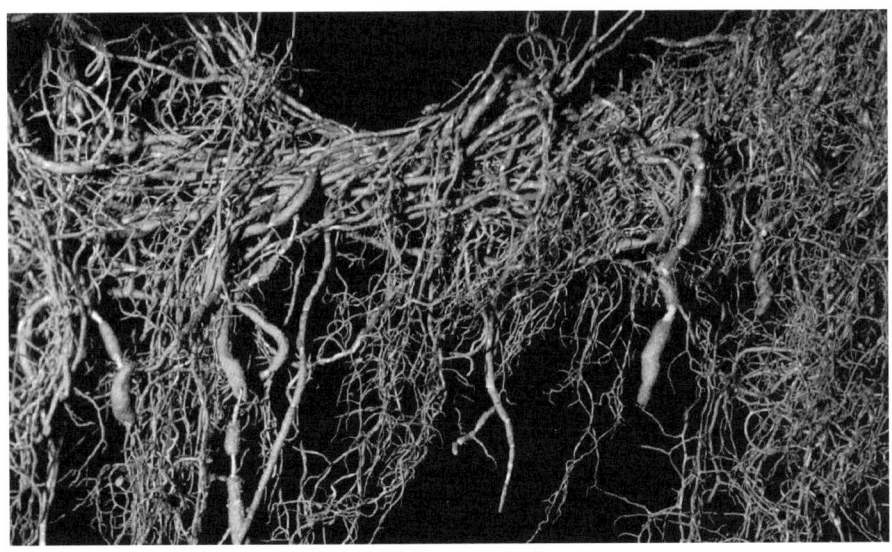

Fig. 37. *Usnea articulata*

There are substantial differences in growth rates between foliose lichens and crustose lichens. Thus, according to one study, the foliose *Parmelia conspersa* has a maximum radial increase of 5.5 mm per year, the placodioid *Diploicia canescens* (pl. 2) an increase of 1.75 mm and the crustose *Rhizocarpon geographicum* (pl. 4) an increase of 0.5 mm. Other studies have shown a growth rate in *Ochrolechia parella* (pl. 3) of 1.5–2 mm per year. It can be assumed that there are variations in successive years, according to differences in rainfall or temperature levels. Besides the constant influences of light, temperature and humidity, growth in lichens is related to the supply of incidental

Fig. 38. *Ramalina menziesii* from California

nutrients, in the form of bird and animal droppings, and to fluctuations in inhibiting pollutants in the atmosphere, water supply or substratum. Besides promoting or retarding growth, an abundance or dearth of such nutrients or pollutants can bring about misshapen or underdeveloped thalli and an absence of fruiting bodies.

Like other living things lichens have a juvenile stage, followed by a period of relatively rapid growth and a final stage of senescence. Growth in the juvenile stage is slow, followed by a steady increase in size until this slows as senescence is reached. The juvenile stage extends from the first discernible establishment of a stable combination of mycobiont and photobiont. The mature stage covers the period during which fruiting bodies or isidia or soralia develop. In the final stage of senescence, decay sets in and accelerates until decomposition is complete. Only older lichens are found in areas where pollution has spread, suggesting that in their early stages young lichens are more vulnerable or possibly find it difficult to become established on an already polluted substratum. Thus, well-developed specimens of *Parmelia caperata* can be found near Fawley Oil Refinery in Hampshire, but small young thalli are absent. Estimates of lichen longevity are particularly hazardous, since the time-scale exceeds that of most other organisms. It is claimed, however, that slow-growing Arctic lichens can survive for hundreds of years and that species of *Parmelia* in temperate zones may live from 10 to 55 years. Crustose lichens are certainly more long-lived. A major difficulty in the discussion of lichen growth and longevity results from the difficulty of determining what constitutes an individual lichen. Very often different lichen thalli of the same species will merge when they touch, making it very difficult to date a specimen by size and growth rate.

The concept of 'death' is difficult to interpret in respect of lichens since individual growths fuse and intertwine and may lack definition. Death and decay of one part of the thallus may be accompanied by fresh growth elsewhere. It is no easier to detect the transition from living to dead tissue. It can, of course, be said that death has taken place when respiration has ceased, but lichens can survive and resuscitate after very long periods of such low metabolism when they seem totally lifeless.

Can lichens, then, achieve immortality? The famous lichenologist, Annie Lorrain Smith, wrote "As a rule – there need be no limit to the age of the lichen plant. There is no vital point or area in the thallus; injury to one part leaves the rest unhurt, and any fragment in growing condition, if it combines both symbionts, can carry on the life of the plant." It is now thought that the previous estimates of thousands of years are probably excessive, but some lichens can certainly live for centuries.

Lichens and metals

A feature of lichens is their capacity to absorb and tolerate heavy metals inside the thallus. Higher plants do not have such tolerance. It does not seem that this has a great physiological value to the lichen, but little is known about its significance. Substrata with traces of lead, copper and other minerals are more favourable to some lichens than to others. For example, the genus *Stereocaulon* is found among the debris of old lead mines and *Acarospora sinopica* is found only on rocks with a high iron content. The metal content that can be endured by lichens can be extraordinarily high and would be destructive to most organisms. Thus *Diploschistes scruposus* has been found with a zinc content of almost 10% of its dry weight, whereas its soil substratum had only 1%. Nevertheless, metal contamination is destructive to most lichens, certainly when constant rainwater drips on to them from copper, barbed wire or zinc (Fig. 40) but even this niche is exploited by some lichens (e.g. *Vezdaea*). Metallic content is not incorporated within lichen cell tissue but is present as minute particles or may be bound to extra-cellular material. The colouration of lichens may often clearly indicate their metallic content. For example, they take on a reddish tinge on rocks with iron sulphide, and copper gives a greenish hue to others which normally have a greyish thallus.

Radioactive substances in lichens

One of the most striking features of lichens is their tolerance to radioactive substances, including strontium and caesium, whether existing naturally or resulting from man-made radiation. This tolerance is such that they can survive and grow where most other forms of life perish and where the continuance of human life is impossible. An important consequence is that lichens in the food chain can place animals and man at risk. Lapps and Eskimos have been found to have higher concentrations of radioactive nuclides in their bodies than other people, owing to their reliance on lichen-eating reindeer and caribou for meat, the radioactive caesium having been concentrated in the fat of the animals. Natural radiation is still the most significant overall factor in creating and maintaining high levels of radioactive substances in humans and animals in some regions of the world but man-created radiation pollution can have disastrous and far-reaching effects. Thirty times the acceptable level of radiation was found in reindeer in northern parts of Scandinavia following the Chernobyl disaster and the effects, even on sheep-pastures in Cumbria, have been long-lasting. One could speculate that, following any worldwide nuclear contamination, lichens might find themselves alone in an almost lifeless world.

Lichens and parasitism

Fungal parasites

The association of fungus and photobiont which constitutes a lichen may be preyed upon by a considerable number of fungal species (about 300 species in the British Isles) that are parasitic by nature. In many cases the parasite fungus feeds on the lichen's photobiont destroying the thallus in whole or in part. This explains the bleaching or browning found in a variety of lichens. A common example is provided by the fungus *Athelia arachnoidea*, which forms light grey patches on *Lecanora conizaeoides* (pl. 2). A number of fungal parasites form galls on the host lichen (e.g. *Guignardia olivieri* on the thallus of *Xanthoria parietina*). In other cases the invading fungus actually replaces the original mycobiont of the lichen and a new and different lichen results. The fruiting body of lichens is also frequently parasitized (e.g. *Arthonia varians* on *Lecanora rupicola* Fig. 39). Some invaders are lichenized fungi that live, at least when young, on or within other lichens (e.g. *Rhizocarpon viridiatum* on *Aspicilia caesiocinerea*).

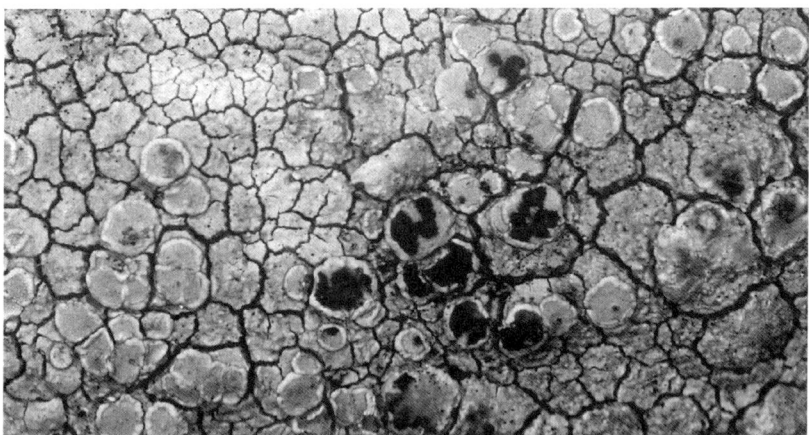

Fig. 39. *Arthonia varians* infecting the fruiting bodies of *Lecanora rupicola*.

Invertebrate parasites

Lichens are attacked by a number of invertebrates including snails, slugs, barklice, psocids, mites, caterpillars and ants. Studies have been made to discover whether the chemical substances produced by lichens can deter invertebrate parasites, but so far results are inconclusive. The likelihood, however, is that they do.

Chapter 6
Lichens and their Environment

The importance of the substratum

A lichen is, with very rare exceptions, firmly attached to, or indeed sometimes embedded in its substratum whether this is rock, bark, soil or anything else, and it is the texture of the substratum that is of the utmost importance. What it must have, above all, is a high degree of permanence and stability. Since lichens are slow-growing, trees with flaking bark, such as planes and gums, may have very little lichen cover, as may crumbling rock surfaces, chalk or loose, friable soils, although there are lichens that can bind the particles of such soils, and sycamore trees often have very good lichen cover. On the other hand, very smooth surfaces of any kind and very hard rocks, such as granite, are unfavourable. It is difficult for a lichen propagule to remain on such rock long enough to become established, although certain species (e.g. *Umbilicaria cylindrica*) manage to grow in fine cracks.

Lichens are a remarkably adaptable life form and can be found on a wide variety of surfaces, including cement, mortar, glass, leather and rubber; thus they are found on concrete paving slabs, tarmac paths, dustbin lids, bonnets of derelict cars and even the carapaces of giant turtles. In view of the current concern with the destruction of natural habitats by man it is worth remembering the vast range of 'unnatural habitats' human beings have created, in the shape of roofs, walls, fences and patios. Man-made structures in lowland areas, where rock is rare, compensate in some measure for the destruction of other, though very different habitats, by draining bogs, tree-felling and the wear and tear caused by ramblers, though some species seem to thrive best on the edges of well-trodden paths.

The situation of the substratum, as well as its nature, is of great significance in that it determines the extent to which the lichen is exposed to, or protected from, sunlight, rain, mist and wind. The direction faced by the micro-site occupied by a lichen is important, enabling it to receive sufficient sunshine or absorb moisture carried by prevailing breezes. Western regions of the British Isles are more favourable to lichen growth than the drier Eastern regions which also receive all the pollution gathered up as the prevailing westerly winds cross the country. Great variations in environmental factors can exist within very small areas, each with its own distinctive 'microclimate'. Even on a medium-sized piece of rock there may be slopes and crevices, or the shading effect of other rocks or trees, that result in quite sharply contrasting conditions of shade, moisture absorption and retention, and nutrient release. On a single gravestone there may be flat, vertical, inclined or indented surfaces, deep shade or full exposure. Decaying trees may release organic material that can be taken up and fences, walls and roofs may collect farmyard and other dust.

An important feature of substrata which determines what species can thrive on them is the degree of acidity or alkalinity, be it in the bark of trees, in limestone and siliceous rocks, in soil, or in other substances. It is largely through the increase in acidity due to atmospheric pollution that 'lichen deserts' are created in and around industrial areas. Alkaline substrata can partly compensate for sulphur dioxide pollution, especially in respect of crustose lichens and air-borne dust from limestone quarries can 'buffer' (i.e. neutralize) acidity in stone, bark and other substratal, thus enabling species sensitive to pollution to survive.

Effect of lichens on their substrata

There has been much discussion of the effect of lichen growth on rock surfaces and on the formation of soil (pedogenesis). Linnaeus saw crustose lichens as the first foundation of vegetation and this view was echoed by Lindsay, Smith and other lichenologists. It was argued that lichens, able to establish themselves on inhospitable bare rock, were also able to collect air-borne or rain-borne particles of organic and mineral origin and make possible soil formation. Subsequently, doubt was cast on this assessment, but the most recent studies have caused a return to earlier judgments. There has also been a recognition that mosses can perform a role as first colonizers and prepare the way for *Cladonia* species. In considering the effect of lichens on substrata, it is important to make a distinction between changes in rock and other surfaces by processes of disintegration (physical) and processes of decomposition (chemical).

Disintegration
Physical invasion of rock surfaces takes the form of penetration by fungal hyphae for up to 15 mm or more but penetration of less than one millimetre is more common. Many crustose lichens are actually embedded in rock, only their fruits being visible on the surface. Presumably the decay and death of such lichens results in a further weakening of their substrata. It has been argued, to the contrary, that a covering of lichens may give protection against rain or wind-borne pollutants and against sand particles or ice crystals which might otherwise erode the rock surface. Crustose lichens are anchored to their substrata by hyphae growing out of the medulla (there being no lower cortex). The medulla serves the lichen as a water reservoir and expands or contracts as a consequence of wetting and drying. The strains that result cause tiny fragments of rock to become detached.

Decomposition
Lichen substances have been shown to decompose the rock surface on both acid and basic substrata. This is especially true in the case of 'endolithic' lichens growing in basic rock. The pits that contained the fruiting bodies of the lichen are often clearly visible. Fungal hyphae first enter very fine cracks in the rock which later expand under the effects of the lichen acids. Similar processes apply in the case of lichens on brick, concrete, cement, mortar and even glass. Lichen hyphae invade these materials and can cause visible damage. On stained glass windows in churches they can distort colours and cause corrosion of the surface of the glass. Removal is only possible at the cost of further damage to the surface of the material attacked.

There is a continuing strongly argued debate as to the degree to which lichens actually cause damage or whether they afford some protection to buildings, monuments and rocks.

Crustose lichens are not always the first lichen colonists. Some foliose lichens, such as species of *Umbilicaria*, may precede them. Higher plants may precede lichens, for example, sand dunes have to be colonized by binding grasses before terricolous lichens have a stable substratum on which they can become established. Species of lichen such as *Icmadophila ericetorum* are important first colonizers and help stabilize freshly exposed peat hags. Thus, besides contributing to soil creation, lichens also have some part to play in its stabilization. If it were not for them and also for algae, mosses and other vegetative cover, weathering resulting from the action of wind and rain would rapidly bring about the erosion of the soil they helped to create.

Damage by corticolous and lignicolous lichens
Rather different considerations emerge in relation to lichens on bark and wood. Lichens on bark may, as on rock, penetrate the upper layers of their substrata by means of medullary hyphae and rhizines. They may also assist the development of insects which may damage a tree by retaining moisture and giving shelter under a foliose thallus. If they block the lenticels or 'pores' of their hosts, they may inhibit its healthy growth. But how far substantial damage results is still in doubt. Nevertheless, it is widely believed that damage is caused, so the practice of spraying fruit trees to destroy lichens as well as other 'pests' is common in some countries. Old and diseased trees often display a luxuriant lichen growth for a time, but this may be a result of their condition rather than its cause. It has been suggested that the decay and disintegration of bark and, indeed, the decay of wood used, for example, in fencing may release substances that provide nutrition and retain moisture thus encouraging lichen development. As the decay of the wood progresses there is a succession of lichen species, each suited to the changing composition of the host.

The importance of atmosphere

The atmosphere is of vital importance for the majority of forms of life. Plants need water and carbon dioxide for the photosynthesis that provides their nutrition. They also draw substances, including water and minerals, from the soil through their roots. For lichens the atmosphere and the mist or rain are the vitally important elements of their environment. This atmosphere carries to the lichen both what is necessary and what is harmful. Whatever the thallus absorbs from a substratum is, moreover, conditioned by the moisture derived from the surrounding air.

One of the most important pollutants that affect lichens is sulphur dioxide. This severely hinders photosynthesis by breaking down chlorophyll and damaging the chloroplasts as well as inhibiting respiration, especially on acidic substrata. Sulphur dioxide is produced in vast quantities by natural processes such as volcanoes but is also a by-product of industrial processes which burn oil and coal and produce sulphur compounds. These enter lichen tissue, either in solution in rainwater or mist, or in gaseous form in the case of sulphur dioxide. Moist conditions allow easier entry into the lichen tissues.

Power stations, smelting plants, factories and domestic fuel consumption concentrated in large cities and industrial areas all contribute to lichen damage. In such cases the level of sulphur dioxide in the atmosphere is due to local pollutants but in others it is due to 'acid rain'. This is caused by air-borne noxious substances being transported often for hundreds of miles and ultimately returning to the earth's surface in rain. The use of high chimneys to disperse gaseous waste may contribute to this happening, the gain in cleaner air at the source of pollution resulting in a corresponding loss at more distant locations; thus it may add to the atmospheric sulphur dioxide produced locally. It is widely suspected that the deterioration and death of trees in the woods and forests of Scandinavia and west Germany is due to acid rain brought by the prevailing winds blowing over these regions from industrial centres as far away as Britain. Conversely, air currents have brought nuclear pollution from Chernobyl, in the Ukraine, to Britain. Recently there has been a fall in the sulphur dioxide levels in Britain due mainly to lower emissions from power stations. Many species of lichen are now returning to habitats from which they have been missing for very many years.

Sulphur dioxide is often associated with other pollutants including nitrous oxide, carbon monoxide and ozone. Fluorides, from industrial plants engaged in the manufacture of ceramics, aluminium and fertilizer products, are less important. Though fluorides are very toxic they are much more local, often killing lichens down-wind of the factory, but the effects diminish within a few miles. Agricultural chemicals are an important and growing source of lichen damage but natural fertilizers may encourage some species. Fumes from motor vehicles have little effect on lichens, since they do not contain sulphur dioxide and the hydrocarbons they do contain cause little damage unless in high concentration with the production of nitrous acid. Roadside lichens do not usually differ from those found elsewhere, save where rock salt is used for de-icing. Since the Clean Air Act soot and smoke do not appear to be a problem, although previously, precipitations accumulating on a lichen could have prevented light from reaching the algae. The alkalinity of soot may indeed neutralize the acidity of some urban substrata.

Fig. 40. Roof showing loss of lichens from rain dripping off a copper wire.

Lichens and their competitors

Man-made pollution apart, lichens are well equipped to cope with environments hostile to most other forms of life. Many are resistant to both drought and to very low and high temperatures, they do not need soil, they can anchor themselves to inhospitable substrata and, within limits, they can adapt to varying levels of light and shade. Although good survivors, they are frequently poor competitors, by reason of their slow growth and relatively small size. On moorland, however, patches of species of *Cladonia* (pl. 1) appear to be able to inhibit the growth of heathland plants and to hold their own. Species of *Peltigera* (pl. 3) can even successfully establish themselves on damp garden lawns, surviving frequent mowing. Mosses may precede lichens by settling in rock crannies or in bark fissures where wind-borne or other humus has accumulated. In such situations, lichens which may also arrive can be rapidly overgrown. However, species of *Ochrolechia* (pl. 15) and *Perusaria* are able to overgrow mosses and produce substances that inhibit the moss from overgrowing them. It has been reported that *Baeomyces rufus* has been found to alternate with mosses, so that a site may be taken over by the lichen for about three years and then be replaced by mosses for a period.

Lichens compete with other lichens. One species may resist the encroachment of another. It is claimed that species of *Rhizocarpon* can develop a 'zone of inhibition' 1–5 mm wide within which other lichens cannot develop. In other cases, one lichen may invade and overgrow another or develop on bare patches in the other's thallus. Alternatively, two thalli may remain separated by the dark borders of their prothalli. In the latter case the result is the patchwork of thalli seen in such species as *Rhizocarpon geographicum* (pl. 4 and Fig. 41 below). On rocks these mosaics may be very stable and last for many years without change. At the same time the balance between the species may be so delicate that a small change in conditions, such as an increase in light levels resulting from the death of a tree, can upset the balance.

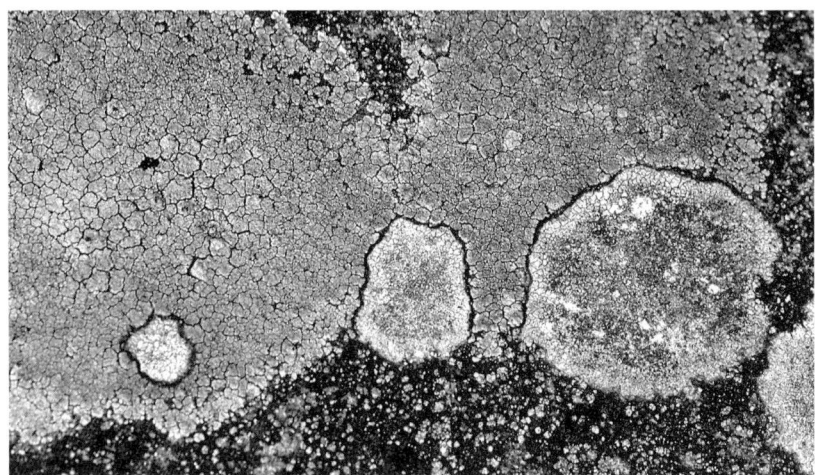

Fig. 41. 'Zone of inhibition' between *Porpidia tuberculosa* (the smaller thalli) and *Lecanora sulphurea*.

Lichen communities

Lichen species, with similar ecological requirements growing together in the same habitat and usually on similar substrata, are described as 'communities' and classified into alliances and associations in the same way as plant communities. Alliances may contain one to many associations as in the relationship between genus and species. The description is based on species frequency in relevés or stands of lichens from typical examples of the community and a type site described. The range of species found in a community is remarkably consistent, allowing the use of shorthand reporting using community, characteristic or faithful species instead of a full list.

The following table should make this clear.

Alliance (suffix 'ion')	Association (suffix 'etum')
Cladonion coniocraeae (two associations)	*Cladonietum cenotae* (*Cladonia* species on acidic tree bases) Dominant species: *Cladonia cenotea, C. furcata* and other moorland *Cladoniae*
	Cladonietum coniocraeae (*Cladonia* species on shaded tree bases, rotting wood, peat) including *C. coniocraea* and *C. macilenta*; it also includes *Lecidea granulosa, Micarea prasina* and *Peltigera* species.
Usneion barbatae (five associations of which two are shown opposite)	*Ramalinetum fastigiatae* (*Ramalina* species on well-lit exposed trees) including *R. canariensis, R. farinacea, R. fastigiata* and *R. fraxinea*
	Usneetum articulato-floridae var. *ceratinae* (on horizontal branches of trees) includes *Usnea articulata, U. ceratina, U. florida,* and *U. rubigenea*

Most communities comprise many alliances and many associations. Those which have served as examples above (*Cladonion conicraeae* and *Usneion barbatae*) are among the smallest. These associations and alliances are associated with definite ecological niches, so that a knowledge of lichen communities can be of value in field work since it gives guidance on what species are to be expected in which locations. For example, *Cladonion conicraeae* is associated with mossy boles or peaty soil, while the *Usneion barbatae* is associated with well-lit twigs and branches in unpolluted areas.

Fig. 42. A well-developed community of the *Usneion barbatae* alliance

Fig. 43. A *Lobarion pulmonariae* community in west Scotland

Classification of of lichens in terms of habitat is necessarily imprecise and dependent on a wide range of ecological and historical factors as well as on subjective observer assessment. Some species may occur in a wide range of communities whereas others are 'faithful' to a specific association. Only an outline of lichen communities can be given here, but a very full guide, on which this section is based, is to be found in a chapter by James, Hawksworth and Rose in Seaward's *Lichen Ecology*.

Corticolous communities
Examples include the already mentioned *Cladonion coniocraeae* and *Usneion barbatae*, as well as communities characteristic of old established woodland. These fall into two main alliances: the *Lobarion pulmonariae*, occurring on well-lit trunks in moist conditions and dominated by large foliose species of *Lobaria*, *Sticta* and *Parmelia;* and the *Lecanactidetum premnae* within the *Calicion hyperelli*, an association dominated by crustose species on the dry side of ancient trunks. Other characteristic crustose associations are found in the *Graphidion scriptae*, including the *Graphidetum scriptae* dominated by lirelliform lichens (species of *Graphis* and *Opegrapha*); and the *Pyrenuletum nitidae*, a community of deep shade in woodlands with species of *Pyrenula* and *Enterographa*. The *Physcietum adscendentis* (dominated by species of *Physcia* and *Xanthoria*) is characteristic of nutrient-enriched bark in the vicinity of farms and intensive agriculture

Saxicolous communities
Saxicolous communities may be found in a wide range of habitats from sea cliffs to mountain tops to man-made buildings or substrata. Although these can be divided into two categories, associated respectively with basic (calcareous) and acid (siliceous) rock, there are many situations where other environmental factors such as nutrients or moisture affect the community.

Basic rock: The most characteristic association of limestone outcrops, buildings and mortar, the *Aspicilion calcareae* includes 114 species, placed in six associations. The first five are *Caloplacetum heppianae* (on exposed, well-lit, dry limestone; *Gyalectetum jenensis* (on shaded, damp limestone); *Direnetum stenhammariae* (on shaded, e.g. north-facing, vertical limestone and buildings); *Leproplacetum chrysodetae* (on shaded limestone underhangs and mortar where it is sheltered from any rain); and *Placynthietum nigri* (on sunny or lightly shaded limestone that remains damp); – this association often includes *Collema* species. The sixth association, the *Xanthorion parietinae*, is widespread on basic rock but is associated with nutrient-enriched sites in both maritime and inland situations such as farm roofs and concrete.

Acid rock: Most of our natural rock surfaces are covered by lichens that fall into visually distinct alliances associated with environmental conditions; shaded (*Leprarion chlorinae*), exposed (*Lecideion tumidae*), nutrient-enriched (*Parmelion conspersae*), and mineral-rich (*Acarosporion sinopicae*). Shaded rock surfaces often support dull-coloured pale to dark species of *Lecidea*, *Lepraria*, *Micarea*, *Opegrapha* and *Racodium* whereas exposed, sunny rocks may support white (*Pertusaria* spp.) or brightly coloured (*Rhizocarpon geographicum*) crusts, as well as foliose species of *Parmelia* (Fig. 44). With increasing nutrient enrichment other species of *Xanthoria*, *Parmelia* and *Physcia* appear, and on mineral-rich rock *Acarospora* and *Stereocaulon* species are common.

Fig. 44. An acid rock community on a grave stone including *Xanthoria* and *Parmelia* species. The inset limestone rose has a different basic rock community containing *Caloplaca* species.

In Britain much of our coastline is dominated by lichen communities, from the black intertidal littoral zone dominated by the *Verrucarietum maurae* to the yellow zone of the *Caloplacetum marinae* to a grey zone dominated by species of *Ramalina* in the *Ramalinetum scopularis*. In areas where gull populations are high the *Xanthorion* is frequent.

Terricolous communities
Lichens growing on soils usually coexist with higher plants which already have designated associations. Four categories of lichen habitat are often used to describe these communities. (Work is continuing to describe coastal and acid soils communities.)

Pebbles: Large or small stabilized pebbles or rocks on exposed sites support the *Huilietum crustulatae*. Included here are *Porpidia* species together with *Rhizocarpon obscuratum* and in slightly more shade, *Baeomyces rufus*. The *Lecideetum erraticae* is found in similar but still more shaded habitats, often in hollows in shingle, and includes *Catillaria chalybeia*, *Buellia* species and *Verrucaria nigrescens*.

Basic soils: Chalk slopes, especially, contain the *Lecideetum watsoniae*. This includes *Protoblastinia*, *Verrucaria* and *Thelidium* species. On flints in this community *Aspicilia* species are common.

Lichen distribution in Britain

Prior to 1982, only 78% of the British Isles had received some form of mapping coverage but since then much time and effort has been spent extending the study of the distribution of lichen species in Great Britain and Ireland. The results from the fieldwork are based on records made of the numbers of species found in nearly 3,000 ten-kilometre grid squares. The results show the relative richness in lichen flora in northern hilly regions and in western and southern England, but the paucity of lichens in the industrial north and Midlands and in the flatter and drier eastern counties, a situation that is rapidly changing as air pollution levels fall and lichen species recolonize areas where they have not been seen since the start of the industrial revolution.

Maps of the distribution of individual species, reproduced from the ongoing fascicles of the *Lichen Atlas of the British Isles* show marked differences. *Lecanora conizaeoides* (pl. 2) is almost everywhere, save in the far north of Scotland, exploiting the gap left by the death from pollution of other lichens. It is, however, now in retreat as general pollution levels fall. *Diploicia canescens* (pl. 2) is only very common south of a line from the Humber to the Wash; *Evernia prunastri* is widespread save in the most polluted areas; and *Verrucaria maura* is, as would be expected of a maritime lichen, strictly confined to the coasts.

The worldwide distribution of lichens

Worldwide, lichens form a substantial part of the biomass. In the arctic tundra lichens comprise over 80% of the living matter, covering several thousand square miles of permafrost. They are so abundant around the world that it has been suggested that, despite their low rate of growth, they make an important contribution to absorbing carbon and helping to delay global warming. Indeed, without their insulating cover the melting permafrost soil could release so much carbon dioxide that this might lead to a runaway warming of the globe.

Substantial and obvious differences exist between animals and plant distributions in the various regions of the world, polar, temperate and tropical. Such differences also exist in respect of lichens, though to a lesser degree. Lichen genera and species are, frequently, universal. It is noteworthy that many lichen species native to Australia are found elsewhere. This is not true of plants found in Australia nor, indeed, of animals. The explanation has been put forward that lichens, being a very ancient life-form, arrived at their present distribution before the continents began to move apart, whereas higher plants and other life-forms entered on their diverse patterns of specialization after the great land movements. The alternative possibility, that lichen spores and particles have travelled and survived huge distances, is unlikely, although they have been collected from the feet of albatrosses caught in New Zealand. Although dispersal on materials (e.g. timber) transported by man may have taken place in a few cases, it would seem that lichens spread and settle in new regions with difficulty, perhaps because of each species' particular needs, in terms of temperature, light and, above all, their sensitivity to substratum and atmosphere. Little or no effort has been made by man to introduce lichens into new habitats, as has happened on a large scale with trees, flowering plants and food plants because of their many and varied uses.

In temperate zones there is substantial similarity between the lichen floras of different regions, certainly greater than that of vascular plants. The lichens of temperate North America are also to be found in Western Europe, such differences that there are corresponding to climatic variation, resulting from the cool summers and mild winters of maritime areas, to the hot summers and cold winters of central continental land masses. Thus, the lichen flora of California (Fig. 45) and that of Normandy have much in common, whereas that of the middle western United States and central and northern Canada has much in common with that of Siberia. Altitude is also a significant factor. There is a distinctive alpine flora that is related to that found in polar regions and which is mainly composed of crustose species such as *Rhizocarpon geographicum* (pl. 4) and sturdy, tufted, strongly anchored species of the genera *Cladonia*, *Stereocaulon* and *Peltigera* which are able to resist cold, violent winds.

Fig. 45. A thick layer of *Cladonia* beside a path in coastal California.

In the tropics, where the heat and high humidity permit the growth of many distinctive plant forms, there is a profusion of *Graphina*, *Pyrenula* and *Thelotrema* species and also of foliicolous lichens (lichens that grow on leaves) and basidiolichens. Lichens which are typical of temperate and alpine regions are also frequent in the hilly districts and high mountains of tropical regions. Over 100 lichen species have been identified in the Sahara, surviving in very dry tropical desert conditions (where higher plants cannot), sheltering under overhangs and exploiting any condensation formed during the cold nights.

Chapter 7
The Amateur Study of Lichens

Where to find lichens

Lichens may be found almost anywhere in the British Isles, from the low tide mark to the tops of mountains. Certain lichens have even been able to exploit the most polluted urban environments. Unlike many groups of organisms, they are also available for study at any time of the year.

It is only possible to give a brief account of some of the more important habitats. Saxicolous lichens are referred to in the previous chapter on communities, together with additional information on corticolous and terricolous lichens.

Corticolous lichens
These are rarely specific to particular species of tree. Some trees, because of the nature of their bark, their low acidity and their retention of moisture, often have a large number of lichen species growing on them. Among these are oak, ash, beech, elm, sycamore and willow. It is not known why, in a row of apparently identical trees, one or two only may be favoured by particular species.

One factor which underlines the need to conserve sites of rare lichens is that, where well-established thalli on old trees in ancient woodlands appear resilient to changes in the environment, little new growth is to be found. This indicates that lichens in their earliest years are particularly vulnerable and once a flourishing colony has disappeared, its re-establishment is unlikely.

Lignicolous lichens
These grow naturally on dead trees and on areas where the bark is missing. The minute, pin-like Caliciales may often be found on the hard wood in the early stages as a tree begins to decay. *Hypocenomyce scalaris* and *Pseudovernia furfuracea* are among those which are frequently found. A more common site for lignicolous species is untreated wood used in fencing and building. Here the succession often runs from *Parmelia* and *Hypogymnia physodes* through to crustose species such as *Micarea denigrata* and *Trapeliopsis flexuosa*. Finally, as the wood becomes more rotten, *Cladonia* species may be found.

Marine–maritime lichens
These are remarkable in that they consist of associations corresponding to more or less clearly defined zones. The lowest of them is the littoral zone between high and low water where *Verrucaria* species blacken the wave-pounded rocks. Above this is the supra-littoral zone of wind-borne spray. Orange species of lichen such as *Caloplaca* are the distinguishing feature here. In the salt-laden air this is an extremely difficult zone in which to survive and lichens are among the few organisms that have succeeded in coping with the extreme stresses of their location including alternating salt and fresh

water. Still higher, towards the cliff tops, where there is less salt spray, grey lichens, including crustose *Lecanora* and fruticose and foliose *Ramalina* and *Parmelia* species, are found. This zone then gradually merges into the non-maritime zone. The depth of the zones, and to an extent the lichen species they support, is effected by the degree of exposure to the waves and also by the illumination of the site. It will be found that limestone shores have fewer species and less zonation than siliceous shores. Seashores, therefore, have their own distinctive lichen flora, not only on rocks but, less conspicuously, on shingle that has become stabilized and even on the shells of limpets and barnacles (Fig. 46).

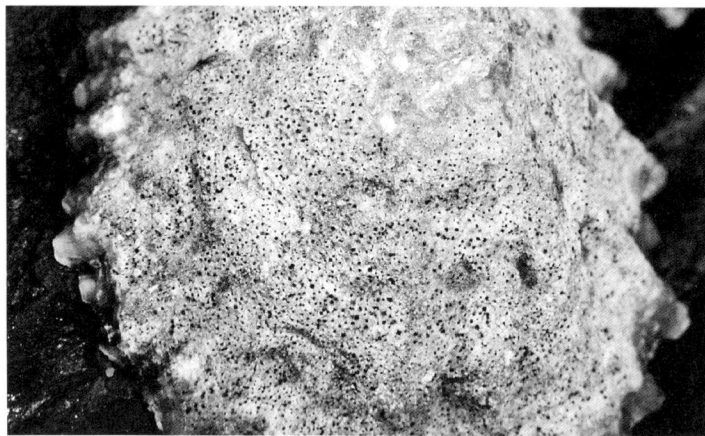

Fig. 46. *Pyrenocollema halodytes* found in the shells of most limpets and barnacles around the coast. Only the small black perithecia are visible on the surface.

Aquatic lichens
These occur in or by the water-line of lakes and streams and consist of species that need, or at least can stand, long or frequent intermittent periods of immersion. They often form zones of species on rocks depending on the normal length of immersion. Those species that are constantly submerged such as *Verrucaria* will only be found in nutrient-poor, usually fast-flowing, water as otherwise they get covered in sediment and are unable to photosynthesize.

Terricolous lichens
These are found on acid soils, peat, coastal soils or sands, and pebbles. Lichens associated with acid soils and peat include *Cladonia* (especially the 'reindeer moss' species) and *Trapeliopsis* species. Moorland can be a rich source of these *Cladonia* species, particularly on bare patches among heather and other plants. Sometimes, large patches are covered with densely woven white and greyish lichens.

A different range of species (e.g. *Bacidia sabuletorum*, *Collema tenax* and other *Cladonia* species such as *C. foliacea* and *C. rangiformis*) are found on basic soils; coastal soils and dunes have a varied population, mainly drawn from calcareous species.

Churchyard lichens

There is probably no better place to learn about lichens than a churchyard. Where else can you find so many species in so small a space, freely accessible at all times of the year and, as likely as not, within walking distance of home! Well over a third of the total species in Britain can be found in churchyards and, in lowland areas where natural outcrops of rock are largely absent, there are no more important places for the conservation of lichens growing on stone.

In England alone, there are over 16,000 Anglican churchyards and they are richly varied in age and setting and structural design. On the other hand, because they are man-made and have arisen as tangible expressions of shared religious beliefs, they have many common characteristics: recurring themes amid the variety. Churches consistently point eastwards towards Jerusalem with their longest walls facing either south, towards the sun, or north, away from it. The sunnier, more exposed surfaces generally carry the most diverse lichen cover and the species are frequently fertile, whereas the shadier north walls tend to be dominated by fewer, often sterile, sorediate or leprose species.

These large expanses of stonework are interspersed with windows and doorways, porches and transepts, decorated with string courses and often strengthened by buttresses. In consequence, a multitude of lichen micro-habitats is created. Individual stone surfaces are variously inclined, variably exposed to the elements and variably shaded and sheltered. One stone may also differ from another in texture, chemistry and age.

Individual stones are, of necessity, bound together by mortars, ideally lime-rich, but at times of almost infinite variety, and the run-off from these may alter the pH of the surfaces below. A run-off more toxic to lichens emanates from metal structures. It may be witnessed, for example, as an iron stain on sills below window grilles even sometimes extending, beard-like, to the walls below. It may also be seen as a copper belt on either side of the lightning conductor. A less visible but even more potent brew is emitted from lead flashings and roofing. Nevertheless some lichens are sufficiently toxitolerant to create a niche for themselves. The edge of the copper belt, for example, is often defined by the powdery lemon-yellow stain of *Caloplaca citrina* (pl. 1).

While some lichens are poisoned by metals, others are sustained by nutrients from bird droppings. You can see the resultant intense orange streakings of *Xanthoria* (pl. 4) even at a distance by standing back and looking up at the parapets, tower ledges and gable ends where the pigeons and jackdaws habitually perch. While taking this longer view, you will almost certainly notice that it is the slopes and horizontals rather than the verticals that are the more richly endowed with a covering of lichens. It is in such places that propagules and the nutrients that sustain them can most easily lodge and where moisture is retained the longest after rain.

The concept of theme and variation extends into the surrounding churchyard. The ancient tombs and the early twentieth century kerbed plots mirror the traditional alignment of the church, while headstones both ancient and modern present their broadest faces most usually to the east or to the west. The oldest memorials are generally hewn and hand-carved from local stone whereas the more recent are likely to be machine-cut, highly polished granites and marbles from much further afield. As far as lichens are concerned, both age and surface texture oblige them to favour the former.

Whatever the age of the memorial stone, there are visibly different communities of lichen on the geologically different rock types. The basic, often pale-coloured oolitic and shelly limestones frequently carry a mosaic of multicoloured crusts on their main vertical surfaces. These may include the chocolate-brown *Verrucaria nigrescens*, pale grey, white-edged circles of *Lecanora campestris*, their centres crowded with tiny, brown, disc-like fruiting bodies and the attractive, orange-lobed plates of *Caloplaca flavescens* (pl. 1). On the tops of these stones you are likely to see the same nitrophilous community you witnessed at a distance on the church roof where the birds perch. Closer inspection reveals them to be mostly leafy lichens, either yellow-orange rosettes of *Xanthoria* (Fig. 44) or pale grey species of the *Physcia* group, perhaps the most typical of which is *Physcia adscendens* (pl. 3), bristling with cilia on its ascending lobe ends. The smoother, sharp-cut tops of marble headstones and crosses, when eventually weathered, support these same lichens, an indication that the stone was also a limestone before metamorphosis.

Sandstones look quite different. They are generally darker in colour (often clearly sandy where the surface patina has been weathered away) and, though variable in pH, are predominantly acidic. Through much of lowland Britain, especially in polluted areas, they carry a more meagre flora. Their east faces and north edges are often highlighted by the almost luminous, sulphur-green, powdery *Psilolechia lucida* and their tops crowned with the mustard-coloured *Candelariella vitellina*, its crust broken up into broccoli-like clumps. Slates and granites are similarly acidic, although, like marble, they too have been reconstituted by pressure or by heat. Whereas, near to the southern or western seaboards, Celtic granite crosses may be dramatically draped with pale green shrubby *Ramalina* species, in central and eastern churchyards low kerbs of the same crystalline stone may need to be combed meticulously with a hand lens before their inconspicuous but distinctive treasures are revealed.

The experienced student of churchyard lichens increasingly has to seek out these less obvious micro-habitats and, because they are often close to the ground, a kneeling mat has become an essential piece of equipment. The chippings within a kerbed grave, whether made of granite, basalt, limestone, marble or even green glass may have lichens growing either directly on them, or in soil pockets and among associated mosses. The more acidic chippings are favoured by cup lichens of the genus *Cladonia* (pl. 1), while the more basic tend to support the gelatinous *Collema* or *Leptogium* species.

In churchyards, lichens also occur on trees, on wooden structures such as benches, on close-cropped grass and pathways and even on rubberized dustbin lids! And no survey is complete without a look at both sides of any boundary walls. Some are so ancient that they may pre-date the church itself.

Such thorough surveys take many hours or even many visits and much satisfaction is derived when a hundred or more species are found at a single site. In Britain, this achievement has now been repeated over sixty times. It is essential that such riches are conserved as well as recorded and this can only be done by wise management. Care needs to be taken when walls are repointed so that disturbance is minimal. Memorials should likewise be left undisturbed and *in situ* wherever possible. It is wise to cut back ivy from church walls and tombstones before it poses a threat and to brush all grass cuttings from low horizontal stonework after mowing. The siting of new shrubs and trees (including millennium yews) needs to be such that important lichen communities are not shaded out in generations to come.

Fig. 47 Members of the British Lichen Society surveying Iwerne Courtney church in Dorset. This church holds the current British record with 170 species.

Woodland and parkland lichens
Lichen species on trees (epiphytes) not only form a significant proportion of our lichen flora especially in lowland areas, but are of particular interest as a most valuable yardstick of the value of a conservation site, indicating the range of biodiversity. Old coppice-woodlands, although often of extreme value for their vascular plant flora, their fungi and animal life, are rarely of great lichenological diversity or value because this traditional method of woodland management inevitably eliminates most of the epiphytic lichen species. The reason that coppice-woodlands, even those known to be ancient, are poor in epiphyte diversity is apparently due to the general lack of mature and/or large trees and the environmental trauma produced by the coppice cycle. During this cycle, heavy shade and high humidity alternate abruptly, after coppice cutting, with full light and very low humidity on the boles of the trees that remain.

A rich epiphyte flora requires the continuity of habitat afforded by stands of mature, if not always ancient, trees together with a combination of good illumination of the tree bole and the humidity which is found in sheltered woodland glades.

Some ancient and famous coppice woodland reserves, such as Bradfield Wood in Suffolk, have *very* limited lichen floras (only 15–20 species in a 1 km square). Open or gladed woodlands, known to be ancient, and with trees of varied ages, may have between 100 and 200 lichen species in a square kilometre. The outstanding examples of these tend to be either little-disturbed woodlands in hilly western oceanic areas with very clean air and moderately high humidity, or else, in the lowlands, are found within the remains of old medieval parkland sites. These, in some cases at least, are now known to have included, at establishment, relict fragments of the original 'wild wood' (Brandon, Ph.D. thesis, *University of London* 1963). Such sites have all been pasture-woodland in the past, with either herds of deer or domestic stock such as cattle (or ponies, as in the New Forest). The animals have kept the woods free of *excessive* regeneration of young trees or shrubs which would shade the tree boles. Nowadays, many such sites no longer have any grazing and are therefore becoming internally denser and much more shaded with a consequent reduction of lichen and bryophyte diversity. However, many fine examples of these pasture-woodlands still remain, even in the lowlands. Examples are Melbury Park in Dorset, Boconnoc Park in Cornwall, Brampton Bryan Park in Herefordshire, Eridge Park in Sussex and the largest and richest of these pasture-woodlands, the New Forest in Hampshire. These can have a good representation of the species of old forest communities (the *Lobarion*). *Very* ancient trees, especially hollow pollards, often have 'pin-head' lichens of genera such as *Calicium* and *Chaenotheca* on dry, old bark or on decorticated wood.

Fig. 48. A moth using *Graphis scripta* as camouflage.

More recently, established or modified parklands often include exotic species among the planted trees.These are usually of less lichenological interest. Ash and sycamore, especially on better-lit boles with a higher, more alkaline bark pH have a special interest of their own. This is particularly the case if the trees are mature or large. Species of *Physcia, Parmelia* (pl. 3) or *Anaptychia* may be common where inorganic fertilizers have not been used but animal excreta provide nutrient sources.

It is really only in these ancient pasture-woodlands that one can still find many species that must have been common and widespread in the 'wild wood'.

Field work

The attention paid to the effects of air pollution on lichens in urban environments can obscure the fact that many species are to be found on garden walls and pavements in city suburbs. The number has been growing steadily since the Clean Air Act of the 1950s. What might be termed 'domestic' lichens include *Lecanora conizaeoides* (pl. 2), *Lecanora dispersa* (pl. 2) and *Xanthoria parietina* (pl. 4). These are widespread save in the most contaminated areas. Small growths of corticolous species such as *Parmelia saxatilis* and *Hypogymnia physodes* may also occur on well-established deciduous trees. Thus, there is no imperative need to embark on special excursions to selected sites nor, indeed, to limit field studies to favoured times of the year. Unlike most other plants, lichens do not wax and wane with the seasons but retain their distinctive form and colour at all times.

It is essential, as well as enjoyable, to carry out some excursions to lichen-rich areas if a reasonably comprehensive acquaintance with a range of lichens is to be made. The ecology of lichens and the total setting within which they are found is vital to their understanding.

Broadly speaking, the most favourable habitats for very many of the species found in Britain are in the track of moist, rain-bearing winds and at altitudes that encourage the formation of mist. Such habitats are found in the hilly regions of North Wales, the Lake District, along the south and west coast and Highlands of Scotland but even in the industrial North and Midlands and on the drier eastern side of England it is usually possible to get into good lichen country without having to travel too far.

For most lichens, sunshine or at least good light is needed and none can survive in dark places, such as caves or the depths of beechwoods. Corticolous lichens are thus best sought either on the edges of woods, or on hedgerow or parkland trees or shrubs. Inspection of oak, ash, willow and elder is often particularly rewarding with some, especially good oak trees, hosting up to 50 different species. It should not be assumed that better specimens will be found away from roads. Indeed, where animals use the roads, there is a likelihood of substratum enrichment from dust from their droppings.

Stone and brick buildings may add an important element to areas lacking trees or natural rock. Man-made artefacts of brick and stone, from menhirs (e.g. Stonehenge) onwards have done much to allow the expansion of the lichen population and the variety of species to be found.

The collection of lichens for study need not be confined to Britain. Collecting abroad lends an added interest to holiday travel and indeed has two advantages. It leads to the discovery of many fascinating lichens unknown in this country and it also makes possible the acquisition of species (e.g. *Roccella*) that are so rare in Britain that collection is out of the question. It is hardly necessary to add that whether at home or abroad collecting should be restricted to locally common species and be kept to a minimum. It should always be remembered that lichen growth is slow and that it takes a long time for losses to be made good but, bearing this in mind, a range of specimens is needed so that home study can be supplemented by microscope investigation and chemical testing. Where identification is known it is better to take a photograph rather than a specimen of the lichen. Photography is a useful adjunct to collection, especially in depicting habitats and in providing illustrations for talks and lectures.

Equipment for field work
The most important piece of equipment needed for field work is a hand lens with a magnification of x8 or x10. This enables the surface features of a lichen to be seen reasonably clearly. It can be carried safely on a cord round the neck so that it is not easily mislaid.

A strong, substantial knife is needed for removing corticolous species and also for cutting out soil terricolous species. This must not be a folding knife unless it is possible to lock it in the open position. A geologist's hammer and a fairly large cold chisel (20–25 cm long) are required for the removal of saxicolous lichens. Length in a chisel enables it to be used for levering away a small piece of rock. Both knife and chisel must be kept sharp if unnecessary labour is to be avoided. Brightly coloured handles can lessen the chances of knife or chisel being lost.

Parts collected should include the edges of the thallus, so that lobes or the hypothallus are collected and, if fertile, they should contain samples of apothecia, perithecia and pycnidia and of isidia and soralia where these are present. Genera such as *Usnea*, *Dermatocarpon* and *Umbilicaria* should include holdfasts because of their value in identification. A portion of the substratum should also be collected to confirm the nature of the rock or bark of which it is composed. It should be emphasized that material collected should be adequate, but no more, for a subsequent full examination to be carried out.

For immediate storage a supply of light envelopes or packets made from scrap paper is adequate. This suffices for most specimens, but small boxes are needed for terricolous species that otherwise would disintegrate under pressure in pockets or bags.

On no account should specimens be left for longer than absolutely necessary in plastic bags unless completely dry as, otherwise, condensation will cause their rapid disintegration. Packets should be marked with details of location, species name, collector, determiner, grid reference and date. Additional information may usefully be added such as results of chemical tests, details of spores examined and Vice-County.

For chemical testing in the field or at home the needs are for a supply of ordinary household bleach that contains calcium hypochlorite (called C in identification books), a small bottle of a solution of 5–15% potassium hydroxide (K) and a small quantity of crystals of paraphenylene-diamine (Pd) with ethanol or methylated spirit in which to dissolve them. These chemicals are offered for sale at all British Lichen Society meetings, details of which may be obtained from the Society's website http://www.argonet.co.uk/users/jmgray/ or from the Secretary at the Natural History Museum, London. All should be used with great care to avoid splashing or ingesting them, especially Pd, which can stain almost any material indelibly and has been thought to be carcinogenic. For this reason, as with dissection, work with chemicals should be carried out on a piece of stout glass. Any material tested must be carefully disposed of. Chemical reactions are best seen if the specimen is placed on a 2.5 cm square of white blotting paper in a petri dish. The red, orange or yellow colours show up much more clearly when absorbed than they do in the specimen itself. The tests must be carried out on the part of the lichens exactly as described in the identification book being used. In many cases it is necessary to test both a fragment of the upper cortex and also expose the medulla (by gently scraping away the cortical surface layer) as the chemical composition of cortex and medulla often vary. It should be noted that C test reactions may only last a second or two before fading. K tests may change colours over about a minute (e.g. yellow turning red) but Pd tests often take some minutes to develop. Other tests used are KC and CK. As the letters suggest, the chemicals are applied in the order given, the liquid from the first test being absorbed on blotting or filter paper and the other chemical then being dropped alongside on the paper so that the chemicals can intermingle.

Home studies

The sooner specimens can be taken from their temporary wrappings and dried the better. For this reason further packages and boxes need to be available on return from field excursions. Drying can be speeded up by exposure to sunlight or by placing on a radiator top, but generally, exposure to room temperature for a period is all that is needed. Dried specimens lose colour but this can be restored to some extent by moistening. Many lichens, unlike most flowering plants, are structurally remarkably hardy and can withstand quite rough handling. As long as specimens are kept dry they are unlikely to be attacked by insects and moulds.

Lichens may be glued on card (portions of both upper and lower surfaces being displayed if this is possible) within their envelopes or boxes. If envelopes are used these can be stored in large cardboard boxes (old shoe boxes are ideal) or in envelope files. For larger rock specimens and for terricolous species small polythene boxes can be used, provided that they are completely dry.

A 'home laboratory' is relatively simple to establish but a dissecting (x20 to x30 magnification) and a compound microscope (magnifying up to at least x400) are desirable. Ideally both should be binocular. If, initially, only one microscope can be afforded the compound microscope should be chosen.

The beginner may be puzzled why two microscopes are necessary. The answer is that, to examine the surface features of any specimen and to dissect sections of fruiting bodies and other features to be more closely studied, an instrument is required that will not magnify too greatly and will enable the viewing of opaque objects. In such a microscope the light is directed on to the specimen from above. To examine spores, fungal hyphae and algal cells which are extremely small and so thin as to be virtually transparent (or capable of being made so by the application of K) high magnifications are necessary and light must be directed onto the specimen from below. Microscope slides may successfully be made from specimens many years old.

Microscopes are not cheap and quality is all important. It can be useful to visit one or two schools or colleges with well-equipped laboratories to find out what instruments are most appropriate. Since the choice to be made is crucial and mistakes are costly, the decision should not be a hurried one. For those who can wait, there is a biennial exhibition of microscopes of all kinds by major manufacturers held under the auspices of The Royal Microscopical Society in London, York, Edinburgh and other cities in turn.

The preparation of specimens for examination under the compound microscope can take two forms. A tiny fragment of the thallus can be detached, wetted, placed on a microscope slide, covered with a cover-slip, and gently crushed by finger-tip pressure. This will reveal hyphal structure and whether green (algal) or blue-green (cyanobacterial) photobiont is present.

A more precise method of studying spores is to take as thin a slice as possible of an apothecium or perithecium. Practice is necessary and a tapering slice producing a very thin edge is best. The best instrument for work on the home bench is the ordinary safety razor blade. If a double-sided blade is used it must be broken in half even before removing it from its packet. This is the only way to avoid accidents. A small piece of thick glass makes an excellent cutting surface. It is important that the tissue examined should be moist. The fragment for dissection should be placed in a drop of water before the section is cut. If successful, details of the form of the asci, as well as the shape, size and number of the spores, will be revealed.

Chapter 8
The Uses of Lichens

By and large lichens have limited major uses. These can be presented under the headings of foodstuffs, medical materials, dyes and perfume manufacture, and decoration. In recent years, their value as environmental monitors has come to be appreciated.

Lichens as foodstuffs

Lichens provide food for reindeer and caribou in arctic and sub-arctic regions where they may constitute as much as half of the winter consumption of these animals. The lichens survive under a thick blanket of snow, but the animals dig down to reach them. The 'reindeer mosses', as they are often called, are mostly species of *Cladonia* and *Cetraria* which form dense carpets covering vast areas. Where there are trees, the corticolous *Bryoria* and *Usnea* species are also consumed by deer. The people of these icy wastes, the Eskimos and the Lapps, store lichen for fodder to supplement the more usual diet of their animals. It is reported that they also produce a flour from 'Iceland Moss' (the species *Cetraria islandica*), from which a form of bread or biscuit is made. The otherwise intensely bitter taste, due to the acid content, is neutralized by the addition of potash. A kind of gruel can be produced by soaking Iceland Moss in water and then cooking it in milk.

Lichens contain very little fat, having only between 1–5% protein but about 90% carbohydrate. This carbohydrate is broken down by deer which have an enzyme 'lichenase' produced in their stomachs. The actual volume consumed must be high, amounting in a reindeer to some 2 kg per day which it obtains from the top 3 cm of the lichen from about 12 square metres of the tundra. Furthermore, lichen pasture must have at least four or five years for its regeneration if grazed by animals and still longer if harvested by man.

Sheep are reported to graze on the soil and rock lichen *Lecanora esculenta*, a native of North African deserts. This lichen is very unusual in that it can become detached from its substratum and blow about on the ground as a light ball of tissue. It is reported to be taken up into the air by the wind, then deposited, and because of this, it is conjectured to have been the manna of the Bible. Lichens are known to have been eaten as a survival food in extreme circumstances by several expeditions.

Species of *Umbilicaria* are eaten in a number of countries as 'rock tripe'. In Japan the lichen *Umbilicaria esculenta* is considered a delicacy and is harvested from the cliff faces and packaged for general sale. A soup is produced from it or else it is served as a constituent of salad. In India a number of lichens are used in salads. Lichens have also been used as a somewhat unsatisfactory substitute for hops in the brewing of beer, and sugar can be extracted from them, though the process is prohibitively expensive.

Medicinal uses of lichens

There is some inconclusive evidence that lichens were used for medicinal purposes in the ancient world. It was only in the fifteenth century that the properties ascribed to herbs began to be codified and a rationale for their use was developed. This was based on the assumption that Divine Providence had chosen to indicate, through the form or 'signature' given to plants, whatever healing properties they possessed. Thus the long, hair-like strands of *Usnea florida* (pl. 4) constituted a sign or 'signature' that this lichen was able to promote the growth of hair. Because *Xanthoria parietina* (pl. 4) was yellow, it was supposed to cure jaundice and *Peltigera canina*, with its long, pointed, and somewhat tooth-like rhizinae, was deemed a cure for rabies.

Lichens were widely used as folk medicines well into the nineteenth century and are still used in some countries today, especially *Cetraria islandica* for coughs. Some lichens contain usnic acid, which is an antibiotic, effective against the tuberculosis bacillus. In recent years the antibiotic property of lichens has been explored and in Germany and Scandinavia a salve, based on usnic acid, has been developed as a means of treating fungal infections. Other lichen products have proved to be effective against certain fungi and viruses harmful to plants. A considerable amount of research is taking place to synthesize these lichen antibiotics for use when bacteria become immune to all those in present use.

Skin contact with lichens may sometimes have harmful effects on people sensitive to them. Some corticolous lichens can cause a form of dermatitis in forestry workers and *Letharia vulpina* can be used as a poison against wolves and foxes. Otherwise, so far as is known, lichens appear to be innocuous.

Dyestuffs from lichens

It seems likely that in ancient times red and purple dyes obtained from *Roccella*, a lichen genus found along the Mediterranean coasts, were used as an inexpensive substitute for the dyes extracted from molluscs to produce the greatly esteemed purple associated with royal and imperial splendour. In the early fourteenth century the use of *Roccella* dyes was re-established in Florence and came to be variously termed 'orseille', 'orchil' or 'icello'. These names were corruptions of the name of the family (the Oricellari) who pioneered their manufacture.

In the eighteenth century another dye of a different origin began to be used in France. This was 'la perelle', which also produced a purple dye. It seems certain that the lichen involved was not *Ochrolechia parella* (pl. 3), as sometimes suggested, but *Pertusaria corallina*.

In Scotland at about the same time another red and purple dye came to be made from *Ochrolechia tartarea* as the principal constituent, and from some other lichens including *Ochrolechia androgyna*, with *Aspicilia calcarea* and *Cladonia pyxidata* (pl. 1) as additional elements. This was termed 'cudbear', the name being a corruption of the maiden name Cuthbert of its patentee's mother. A brown dye was produced mainly from *Parmelia omphalodes*. 'Crottle' is a term used to denote the dye-producing lichens in Scotland.

Lichen dyes are sometimes still used for Harris Tweed but otherwise they have been superseded by much more easily produced and more permanent synthetic dyes. Occasionally efforts are made to revive the craft of crottle dyeing, but they are vigorously resisted by conservationists because of the vast quantities of lichen material needed. This may be the same, in weight, as the wool itself to which it is applied. A lichen dye is the vital ingredient in the making of litmus paper used in school laboratories.

Lichens in the perfumery industry

The part played by lichens in the perfume industry is still appreciable, it being estimated that 9,000 tonnes are used each year. The two lichens of major importance are *Evernia prunastri* (oak-moss) and *Pseudevernia furfuracea* (tree-moss). The lichens, once gathered, are stored for a time, as this seems to develop their distinctive odour. Their main function, however, is to 'fix' other perfumes, which otherwise would lose their special character. In his book, *The Vanishing Lichens*, Richardson gives a fascinating account of lichens in the perfumery industry, although he found reluctance on the part of manufacturers to enter into discussion on exactly how much is used. Perfume manufacture involving lichens takes place in many European countries and especially in France, Italy and Germany. Lichen material is largely gathered in southern France, Morocco and the former Yugoslavia.

Lichens as decorative materials

A common use of lichens is in creating decorations, wreaths and similar displays. These include their use in vase arrangements with other plants and also in depicting human and animal forms in tableaux for festivals and exhibitions. Lichens are also used to represent miniature vegetation in models of railways and buildings and in commercial architect presentations of their plans for the surrounds to buildings. Lichens, especially species of *Cladonia*, are used by florists for packing delicate flowers. A study made of the harvesting of a single *Cladonia* species (*Cladonia stellaris*) for this purpose in Finland showed that about 18,000 tonnes were exported every year to the former West Germany and other European countries, in addition to what was used in Scandinavian countries.

Scientific uses of lichens

Apart from the uses of lichens outlined, there are others of scientific importance. The first is that they are useful monitors of environmental pollution, the second, that they serve as indicators of the minerals to be found in the rocks on which they grow, and the third, that they can be used to give a good indication the age of rocks and the man-made structures on which they grow.

The growth or absence of lichens cannot compete with refined scientific instruments and chemical testing as exact measurers of pollution, but they do give a generalized picture of the consequences of excess sulphur dioxide in a given location. More important, perhaps, is that the presence or absence of variously sensitive indicator species shows visually, even dramatically, the difference between polluted and unpolluted stretches of countryside and, by observations taken over a period of time, the extent to which pollution has increased or diminished. Their value as indicators of past levels of pollution is considerable. Crustose lichens on rock, stone or brick are less revealing since they are not as exposed, to the same extent, to atmospheric pollution and the substratum can have a modifying effect.

The sensitivity of lichens to pollution varies from species to species to a marked extent so that it is possible to map concentric zones, graded in terms of levels of sulphur dioxide in the atmosphere. So far the mapping has been attempted only in respect of corticolous species which are the most sensitive to atmospheric conditions. Studies published in 1970 showed that lichens could be divided into ten zones of pollution sensitivity, and also where the corresponding zones were located on an outline map of England. In the first zone no lichen growth of any kind is to be found. Then follows a zone with only a form of alga. Subsequently *Hypogymnia physodes* and *Parmelia caperata* (now rapidly returning to many areas where it has been absent for many years) make their appearance, followed by zones with lichens of increasing sensitivity including such species as *Usnea subfloridana* and *Pertusaria hemispherica*. Finally, in the outer area with the cleanest air come such infrequently encountered species as *Usnea articulata* and *Teleoschistes flavicans*. Illustrations by Clare Dalby have been published grading lichens in a similar way.

There has been considerable change in pollution levels in urban areas during the past two decades and there is need for pollution and distribution mapping to be constantly revised. Indications are that the reduction of air pollution due to the fall in sulphur dioxide levels by some 50% between 1960 and 1980, and by a further 60% between 1980 and 1987, has resulted in quite startling improvements. A study of 50 sites in north-west London in 1988 revealed that, of 49 species recorded, 34 had not been found in an earlier study in 1970 and, more surprisingly still, eight had not been seen since the late eighteenth century. Foliose and fruticose lichens are most sensitive to air pollution. Only one such species was found to have survived in 1970, and then on only one site. This was *Hypogymnia physodes*, which now exists on no fewer than 34 sites. By 1980 *Evernia prunastri* was found on eight sites but on seventeen in 1988. These newly re-appeared lichens are necessarily small and certainly sensitive, but their numbers suggest that a trend has been established.

Signs of pollution damage in lichens are the fading of their normal colour, patchiness in the centre of the thallus and a peeling off from their substratum. Those thalli that survive are stunted and do not develop fruiting bodies. Foliose and fruticose species are most severely affected because of their greater exposure to adverse atmospheric conditions. Many crustose saxicolous species are substantially resistant. These include *Candelariella vitellina*, *Lecanora dispersa* (pl. 2) and *Lecanora muralis* (Fig. 49). These species are found on basic rocks that neutralize the acidity of the atmosphere to which they are subjected. It will be realized that pollution, save where heavy and varied, can act selectively. An invasion of limestone dust can spell the end of acid-loving species and, as stated earlier, a trace of lead in the atmosphere can favour others. *Lecanora conizaeoides* (pl. 2), which spread in this country as this form of pollution progressed, actually thrives in conditions of sulphur dioxide pollution. With the cleaner air of recent years it is now becoming less common.

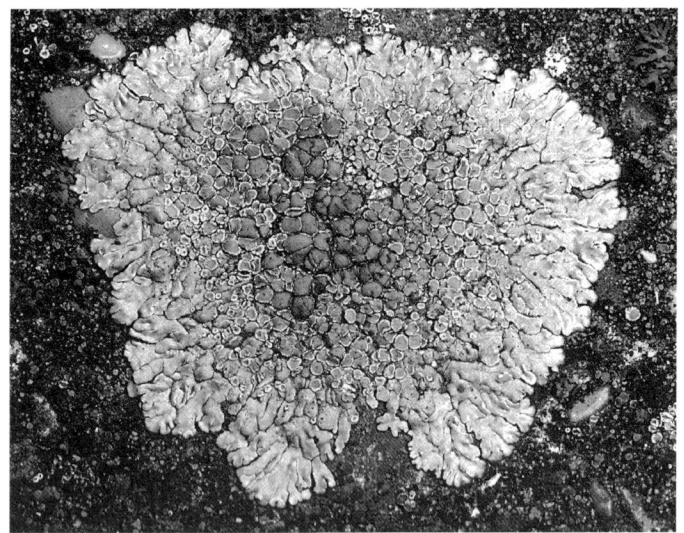
Fig. 49. *Lecanora muralis.*
A species that has spread onto man-made surfaces in semi-polluted areas.

As indicators of minerals lichens can be of service since they can concentrate within their thallus a variety of metals, including zinc, cadmium, iron and copper, their accumulation resulting from the slow growth, longevity and high tolerance levels of many species. Here again, they cannot rival modern methods of metal detection, but they can supplement them, in that a survey of lichen cover in a site of possible interest is a rapid and economical way of making a first study in regions where minerals are dispersed rather than concentrated in seams or veins. They have their uses, too, as sensors of radioactivity from man-made sources.

An important and interesting use of lichens is in dating the substratum on which they are found. To do this it is necessary to select a suitable species to be used as a monitor and then determine its rate of growth. Old stone surfaces, of which the age is known, such as church walls and gravestones, are invaluable in establishing a baseline. Lichen size, on the surfaces of which the age is to be established, are then measured and together with their growth rate plotted on a graph. This technique, known as 'lichenometry', is especially useful for dating surfaces exposed to lichen growth during the past few hundred years and thus can help in ascertaining the age of buildings or the movement of glaciers. Only certain lichens are suitable for use in lichenometry, notably those that have clearly-defined thalline margins and are long-lived. Among these the genus *Rhizocarpon* (pl. 4) is favoured, as it meets these criteria and is also very plentiful, at least in upland areas, but various problems have to be solved. One is the selection of representative sites within the area to be studied as even small changes in conditions may affect growth rates. Another is whether to take measurements of the largest thallus, which may in fact be several thalli which have merged, or base them on a number of smaller thalli. These problems, and the promise that the technique holds, have contributed to the fascination of lichenometry felt by those who have used it as a research tool.

Chapter 9
The Classification of Lichens

The history of lichen classification

The term 'lichen' is from the Greek and means an eruption, wart or even leprosy. It was first used by Theophrastus in the 3rd century B.C. in his *History of Plants* to describe growths on the bark of olive trees. But little further mention of such growths is to be found before the herbals of the fifteenth and sixteenth centuries, which signalled the coming of a systematic interest in the medicinal uses of plants. In the herbals, lichens were placed among the mosses. Among those particularly noted were the conspicuous 'tree mosses', the foliose *Lobaria pulmonaria* (pl. 3) and the equally conspicuous pendant species of *Usnea* (pl. 4).

Towards the end of the seventeenth century the work of the Italian, Marcello Malpighi (1628–1694) and of the Scot, Robert Morison (1620–1683) marked the beginnings of close observation and of the skilful drawing of lichen specimens. Malpighi made significant contributions to the study of plant anatomy and structure as a result of his great skill and persistence in the use of the primitive microscopes of his time. He was a pioneer in the study of cell structure in plants and his interest in fine detail led him to examine and describe soredia, which he took to be seeds. Like Malpighi, Morison was trained in medicine (which then embraced all that now comes under the title biological sciences) but his main interest was in plants. He became the first professor of Botany at Oxford in 1669. His originality lay in recognizing that there was a fungal component in lichens although, like his predecessors, he still associated them with mosses and described them as 'musco-fungi'.

Up to this time all plants had been treated and named as individual species and were only loosely associated in somewhat arbitrary groupings. But a decisive step, and one with far-reaching consequences for lichen classification and taxonomy, was taken by the Frenchman, Joseph Pitton de Tournefort (1656–1708). In his *Elémens de botanique* (1694) he formulated the vitally important concept of genus as a 'cluster of species' with common characteristics. He also separated lichens from mosses and set them aside in their own genus, 'lichen'. He was, therefore, the founder of modern taxonomy and he provided a conceptual basis for lichens to be treated as a distinctive group in subsequent classificatory systems.

Some thirty years later, in his *Nova plantarum genera* (1729) the Italian botanist, Piers Antonio Micheli (1679–1737) made a first comprehensive survey of the thallophytes, including fungi, lichens, liverworts and mosses in his overview. Like Malpighi earlier, he used microscopes in his studies and with them led the way in the examination of spores and mycelia. He refined the concept of the single genus to which lichens were still ascribed by splitting this into 38 orders, in accordance with thallus form and location of apothecia and soralia. Because of his original work in the study of fungal anatomy Micheli can be regarded as the initiator of mycology as a separate branch of botany. The long-standing association of lichens with mosses was, however, still continued in the work of Johann Jacob Dillenius (1687–1747), a German physician who was a successor of Morison in the Oxford chair. He was an enthusiastic student of cryptogams and in 1742 his major work, *Historia muscorum*, presented a new classificatory system for the lower plants. He split the then known lichens among three genera, *Usnea*, *Coralloides* and *Lichenoides*. Nevertheless, as the title of his book indicates, he still regarded lichens as akin to mosses, and his approach was limited in that he concentrated his attention on thallus form to the exclusion of other features. Dillenius built up an excellent herbarium collection, which included lichens.

The number of lichen species recorded by botanists in the mid-eighteenth century varied greatly: Micheli recorded 298, the Swiss botanist Haller 160 and Linnaeus only 80.

Linnaeus and lichens

The great Carl Linnaeus (1707–1778) showed no great enthusiasm for lichens, which he termed the 'rustici pauperrimi' ('the poor little peasants of nature'). He reverted to placing them in a single genus, though he divided this into seven sections (e.g. foliose or 'Foliacei', filamentous or 'Filamentosi').

The major contribution of Linnaeus to biology was, of course, his all-embracing *Systema Naturae* (1735) (and, later in 1753 for botany, *Species plantarum*), in which he presented a system of classification for the whole of creation. This was divided into three great Kingdoms: animal, vegetable and mineral, a division becoming so familiar as to feature in parlour games and quizzes. He took as the basis of his classification of plants their sexual characteristics, notably as revealed in their stamens and pistils. This was a new and original approach to plant taxonomy but it could not accommodate the cryptogams in general nor the lichens in particular. But the clarity that the Linnaean system gave to the concept of a species was of great significance generally.

Besides laying down the principles upon which classification should be based, Linnaeus prescribed an orderly system of nomenclature. This was based on the giving of a generic and a specific name to each plant. Other writers had done this occasionally, but it had been more usual to adopt descriptive names, some of cumbersome complexity. For example, *Rhizocarpon geographicum* (pl. 4) was, in Dillenius, *Lichenoides nigro-flavum, tabulae geographicae instar pictum*, and *Haematomma ventosum* was *Lichenoides tartareum lividum, scutellis rufis, margine exili*. Besides requiring that each plant should have a name incorporating only its genus and species, Linnaeus prescribed (and this became and has remained universally accepted practice) that every species had to be defined in a short diagnosis, to be supplemented by a longer and more detailed description. By diagnosis is meant a brief, terse note of the essential characters of a plant that distinguishes it from others and by description is meant a detailed statement of all the features of a plant, presented in a systematic and orderly fashion.

Acharius and the beginnings of modern lichen taxonomy

It was with Erik Acharius (a fellow countryman of Linnaeus, and indeed one of his students and last disciple at Uppsala), that the study of lichens entered a new stage. Acharius was unique in that he devoted almost all his leisure time to lichens and to describing and classifying them. As a result he presented in his *Methodus Lichenum*, 1803 and his *Lichenographia Universalis*, 1810, a taxonomic structure of orders, genera and species that became the basis of all succeeding lichen systematics. He also presented diagnoses of his taxa, following Linnaeus' prescription. In his chosen field he went far beyond his master, for Linnaeus had concerned himself almost exclusively with the higher plants. Acharius sought to bring order to the study of cryptogams. Many terms which he introduced, including 'thallus', 'podetium', 'apothecium', 'perithecium', 'soredium' and 'cyphella', still persist.

Acharius can also be acclaimed as the first great lichen enthusiast if the story told of his death can be believed. This was that he was so overcome with joy at receiving a collection of lichens from Spain that he died within a few days. A more mundane explanation is that he had a fit of apoplexy, unrelated to his pleasant surprise.

The distinctiveness that Acharius ascribed to lichens as a group, despite his unawareness of their true nature, set them aside from the other fungi, although he sometimes confused what are now termed lichenized and non-lichenized fungi. From his time, lichenology as a study in its own right, with its own literature, specialized vocabulary, herbaria and professional and amateur devotees, became established on a world-wide basis. Only in recent years have efforts begun to integrate it more closely with the wider field of mycology.

Following a workable system for the identification and classification of lichens being established by Acharius, it was possible for advances to be made in two directions, towards detailed studies of individual genera, such as *Cladonia* or *Sticta,* and towards the scientific analysis of the internal structure and physiology of lichens.

Fig. 49. Erick Acharius 1757–1819

Progress in systematic identification helped to promote the inclusion of lichens among the many targets of the builders of botanical collections. These came to include not only native lichens, but also others collected from many parts of the world.

Mid-nineteenth century emphasis on fruiting bodies

With the development of more effective optics and the use by botanists of the early compound microscopes, attention was increasingly concentrated upon lichen spores as the most significant means of identification. The leader here was the Italian botanist, De Notaris, who noted the different shapes, sizes and colours of the lichen spores he examined. He was followed by others including Tulasne in France and, in Britain, Lauder Lindsay, William Mudd and W.A. Leighton. Lindsay's *Popular History of British Lichens* (1856) was the first of its kind. It is delightfully written in a vivid and charmingly ornate style and has attractive colour plates of lichens and of details of their structure. Mudd's *Manual of British Lichens* (1861) has an interesting appendix of illustrations of over a hundred spore types; and Leighton's *Lichen Flora of Great Britain, Ireland and the Channel Islands* (1871) was for many years the leading lichen flora of this country. It was not in fact superseded until Annie Lorrain Smith published her extensive *Monograph of the British Lichens* in 1911.

Mention should also be made of the voluminous works of William Nylander, a Finn who, after early work in entomology, devoted himself almost entirely to lichens and, in particular, to their identification and classification. He was an enormously industrious collector, not only of European lichens, but also of the lichens of Japan, New Zealand, India and Ceylon. But he can be said to have hindered progress in lichen taxonomy by his over-eagerness to establish new species. On the other hand he made a significant contribution to techniques of lichen identification by introducing the use of potassium hydroxide (K) and calcium hypochlorite (C).

The importance of the microscope

The microscope has a special significance in the history of lichenology, since major advance depended upon the examination of cell structure, spores and ascocarp development. It was not until the end of the sixteenth century that the combination of lenses to produce a magnified image had been achieved and the discoveries of Malpighi become possible. In the following century Richard Hooker, the first major English student of microscopy, devised much more sophisticated instruments and used them to examine, among other things, the cells of cork, employing the term 'cell' as a biological concept for the first time. By the beginning of the eighteenth century it became a matter of pride to own a microscope and to be able to use it to display features of plants and insects to others. In fact, the microscope was the plaything of the dilettante rather than an aid to scientific study. All this changed after the discovery of achromatic lenses in the 1830s. These obviated the blurring and false colours of earlier instruments and made the study of fine cellular structures in animals and plants at last possible.

The present state of lichen classification – lichens as 'lichenized fungi'

The definitive work on lichen classification, the huge ten volume *Catalogus Lichenum Universalis* of Alexander Zahlbruckner (1860–1933) has only recently been supplanted by more modern works. This work was published over a twenty-year period from 1921 and hence was, in part, posthumous. The *Catalogus* has since been supplemented by the *Index Nominum Lichenum* by I.M. Lamb, published in 1963. 'Zahlbruckner' and 'Lamb' are still indispensable reference works for all lichen herbaria and literature, including that of the Natural History Museum, but they are regarded with mixed feelings by most lichenologists. On the one hand they together constitute the most complete historical record of lichen names; on the other hand they continue the long-established tradition of viewing lichens as a separate and distinct group of organisms, set apart from the fungi. This tradition is now keenly contested. Present-day lichenologists integrate lichens with the other fungi. The argument is that lichens result from a range of diverse forms of fungus which are 'lichenized' by being brought together with their algal or cyanobacterial partners. Against this, it must be pointed out that many lichens have characteristics not possessed by either fungi, algae or cyanobacteria, namely soredia, isidia and many distinct 'lichen substances' (these are produced by the fungal partner). Lichens in their ecology have more affinities with the mosses than with many fungi.

A major task facing lichen taxonomy now and in the foreseeable future is accepted to be the integration of lichens within the ascomycetes. This is made difficult, however, by the classificatory system for the non-lichenized fungi not adapting well to the incorporation of lichenized fungi. An early attempt at the integration, made in the late nineteenth century by the Finnish lichenologist, Vainio, was nullified by the publication of Zahlbruckner's overwhelming work.

In recent years the 'integrationists' have been helped by the interest now being shown by mycologists in ascocarp structure and the development of inter-ascal tissue in the ascomycetes. It is here that the closest affinities are being revealed between the lichenized and the non-lichenized fungi. Criteria are becoming available for the revision of accepted genera and species, the determination of which has hitherto depended on the external characteristics of thalli, apothecia, perithecia, isidia, soralia and spores. The concern with fine structure has been made possible by electron microscopes with powers of magnification far in excess of those of the most effective light microscopes. Technical advance of another kind has been made by the development of thin-layer chromatography, which has permitted the detailed study of the chemical composition of lichens and especially of their unique 'lichen substances'.

Attention to fine structure and to chemical characters has substantial advantages for taxonomic accuracy, since it reduces 'interference' from ecological factors. Morphological characters of thalli and of fruiting bodies are greatly influenced by accidents of habitat as expressed through exposure to light, moisture and pollution. Nevertheless, morphology will continue to have a major role to play in taxonomy. Cytological and DNA information and chemical characters extend across genera and species, and a taxonomic system cannot be based on them alone. Moreover, a taxonomy making full use of morphological information has enormous advantages for the field worker.

The naming of lichens

The naming of lichens, as of all living organisms, is of the greatest importance, since endless confusion could result if scientists had to struggle with a multiplicity of popular names in a multiplicity of languages. Fortunately, it has been possible to achieve worldwide agreement not only that Latinized names should be current in all cases, but that there should be detailed rules governing biological nomenclature. These were, in respect of plants, first brought together in the International Code of Botanical Nomenclature, first adopted in 1867, and revised at periodical International Botanical Congresses. The Code is a fascinating document, which makes clear the difficulties in keeping track of tens of thousands of plant names and changes in them. It lays down:

1. That any plant must have, as Linnaeus prescribed, only one name, consisting of both the genus name and the species name. It is important to remember that, in naming a lichen, it is only the fungal partner to which the name strictly applies.
2. That the name must be a Latin name or have a latinized form as when a person's name is incorporated (e.g. *Lecanora jamesii*).
3. That the only correct name for a plant is the earliest name that has been validly published and follows the rules, although it is possible for a well known, but later name to be conserved, and confused names to be rejected. Valid publication means that the name has been published in print, is made generally available (e.g. in a book or journal) and is accompanied by a description, and has its application fixed by means of a type specimen.
4. That the type specimen should be carefully preserved, for example, in a recognised depository and be acknowledged to be the type specimen. This does not mean that the specimen is 'typical' in the usual sense of the term, but only that it is the specimen to which the name applied. If the specimen is lost or destroyed, provision is made for a substitute to be accepted. Some of the names used for these type specimens are as follows:

   Epitype   – A specimen or illustration used for features missing on the holotype.
   Holotype   – the specimen chosen by the describing author as the 'type' for the name being described.
   Isotype   – a duplicate of the holotype. Distributed when there is sufficient material of the holotype to allow this.
   Kleptotype – a piece broken off from the holotype (often originally stolen!).
   Lectotype   – a piece selected later from the original material when no holotype had originally been chosen.
   Neotype   – a specimen chosen later as the type specimen when the original material is missing.
   Paratype   – Any specimen other than the holotype on which a new account of a species or group is based.
   Topotype   – a specimen collected at the same site as the holotype.
   Typotype   – The specimen used when the type of a name is an illustration.

Valid naming is accepted as having begun in 1753, the year in which Linnaeus published his *Species Plantarum*. Names devised and published before that date have been remodelled as necessary. Since the Code was adopted there have been constant efforts made to assert the priority rule and thus avoid any confusing re-naming of plants and to establish which one, of all the names by which a plant may have been known, is both valid and legitimate. To be legitimate a name must be in a proper form and must not, of course, have been assigned to another plant.

The significance of the Code is enormous. Its authority is accepted throughout the world despite or perhaps because of its language being derived from Latin. Botanists cannot be compelled to follow its rules in any legal sense, but all botanical institutions and publications accept its precepts as they would those of a major dictionary. There is constant discussion of its interpretation in cases of dispute and much scholarly effort is expended to establish which names are acceptable. Papers frequently appear which discuss whether herbaria specimens have been correctly named. Changes in names must follow when new knowledge, perhaps following an advance in chemical techniques or in the use of the electron microscope, demands the transfer of a species of one genus to another or the introduction of a new genus. Taxonomy, indeed, is a reflection of the changing state of knowledge and not a statement of immutable relationships.

An interesting feature of biological nomenclature is that the name of the author (that is, the person giving a plant or other biological entity its name) can be stated, as in *Lecanora expallens* Ach. The author's name, Acharius, is abbreviated to Ach.

When there is a change of genus, this is recorded by the name of the revising author being added: *Micarea confusula* (Nyl.) Hedl. The original author's name is Nylander, abbreviated to Nyl. He named the species *Lecidea confusula* but Hedland, abbreviated to Hedl., placed it subsequently in the genus *Micarea* and his findings received a sufficient degree of acceptance for the change to be accepted.

Other formulas include the terms, 'ex', 'non', 'sensu-non'. Examples are: *Schismatomma decolorans* (Turner & Borrer ex Sm.) Clauz. & Vezda. The original joint authors were Turner and Borrer, but it was Smith who was the first to validly publish the name *Lepraria decolorans*. Subsequently, Clauzade and Vezda placed the species in the genus *Schismatomma*. The uses of 'non' are now largely obsolete.

Arthopyrenia rhyponta (Ach.) Massal. non Borrer. This means that the species referred to was described by Acharius and revised by Massalongo and not another species to which the same name had been given previously by Borrer.

Lecanora impudens sensu P. James non Degel. This means that the name *Lecanora impudens* is used as P. James used it and not as Degelius did.

Auctorum non (auct. non.) is used where a citation uses a name that has been misapplied. As in *Nephroma laevigatum* auct. non Ach. meaning not as identified by Acharius.

The extreme care taken in the scientific naming of plants and other life forms is not only to ensure that each species can be identified without any confusion, but also to arrange each species within an overall hierarchical system. Thus, beginning with the species, there is an ascending order of generality. Like species are grouped as a genus, like genera as a family, like families as an order, like orders as a class, like classes as a phylum and all phyla together form the regnum or Kingdom. Traditionally fungi have been placed in the plant kingdom, but the present practice is to place them, whether lichenized or not, in a Kingdom of their own, as one of the five or more kingdoms now recognised by taxonomists.

The species is the basic unit of the taxonomic pyramid. But the determination of a lichen species presents difficulties not found in respect of most plants. Although propagation has been achieved under laboratory conditions this is very difficult and long and sustained experiments in lichen cultivation cannot be entertained. Therefore, the taxonomist has to decide whether or not the term 'species' applied to any lichen is justified by considering its thallus form, reproductive structures, anatomy and chemical composition. The task is made harder by the considerable variations in lichens that result from ecological factors (light intensity, temperature, moisture, atmospheric pollution). Such variations can be accommodated by the use of the ranks of 'subspecies' and, when minor, of 'variety' or 'form'. This approach is not, however, in favour by many lichenologists.

Unlike most other groups, such as flowering plants, the higher levels of the taxonomic pyramid are not much used in basic identification due to the great variety in the groupings. For example the family *Teloschistaceae* contains genera with most of the different growth forms: leprose, crustose, placodioid, foliose and fruticose.

The various taxonomic levels of a typical lichen e.g *Lecanora rupicola* var *efflorens* are as follows:

Kingdom	*Fungi*
Division (phylum)	*Ascomycota*
Subdivision (subphylum)	*Pezizomycotina*
Class	*Lecanoromycetes*
Order	*Lecanorales*
Suborder	*Lecanorineae*
Family	*Lecanoraceae*
Genus	*Lecanora*
species	*rupicola*
variety	*efflorens*

Chapter 10
The Literature of Lichenology

There are two books currently available which provide keys for the identification of the lichens of Great Britain and Ireland. These are:

Dobson, Frank, *Lichens: An Illustrated Guide to the British and Irish Species*, The Richmond Publishing Co., 4th edn., 2000.

Purvis, O.W. et al. (eds.), *The Lichen Flora of Great Britain and Ireland*, The British Lichen Society in association with The Natural History Museum, 1992.

Dobson's book is the easier to use for beginners, as virtually every species, of which over 600 are treated, is illustrated by a colour or black and white photograph and has a thumbnail map showing where it is to be found. In addition there are line drawings of spores and other features. The keys are clearly set out and easy to follow.

The *Lichen Flora* is a very substantial and detailed work, the result of many years' work by major lichenologists in Britain and overseas colleagues. It is the only work of its kind to appear for nearly eighty years. It is illustrated only by line drawings of morphological details of lichens, but it has very full descriptions of all the 1,730 species with which it deals. The keys, whilst necessarily not as quickly mastered as those in Dobson, are nevertheless user-friendly.

An older and popular guide, with numerous illustrations, is *The Observer's Book of Lichens*, by Kenneth Alvin, Warne, 1977. This pocket-sized book deals with 170 species. It is unfortunately now out-of print (OOP), but second-hand copies can sometimes be found.

Another illustrated guide is *The Oxford Book of Flowerless Plants*, illustrations by B.L. Nicholson, text by Frank Brightman, Oxford University Press, 1966 (OOP). This is distinctive in that it groups lichens together with ferns, fungi, mosses, liverworts and seaweeds according to habitat (seashore, grasslands, upland areas, wet places and woodlands).

Finally, among visual aids to lichen identification there are two folding charts, with information and keys by Frank Dobson. On the reverse of these keys on *Lichens and Air Pollution* and *Lichens on Rocky Shores* are the exquisitely executed paintings by Claire Dalby, each covering some 50 or more species. They are laminated sheets for field use that fold down to A5 size. These are produced by The Richmond Publishing Co Ltd. Also available is a set of five A4 sheets on the identification of churchyard lichens. The notes are by Tom Chester and the reverse of each sheet has 12 colour photographs of churchyard lichens. They are obtainable from The British Lichen Society.

To check the distribution of lichen species in Britain it is necessary to consult the *Atlas of the Lichens of the British Isles,* This is an ongoing programme by the British Lichen Society to produce loose-leaf A4 fascicles of distribution maps of British and Irish lichens. Some of the genera have introductions and keys. The reverse of each map contains the latest information on that species.

Lichen names, for reasons already discussed, are constantly being revised. Checklists are essential reference tools. An important work of this kind is the *Checklist of Lichens of Great Britain and Ireland,* by O.W. Purvis, B.J. Coppins and P.W. James, published by the British Lichen Society, 1994. Another checklist which places most lichens among the fungi which have asci, thus reflecting the current movement to bring together lichens and fungi is *The British Ascomycotina: An Annotated Checklist,* by P.F. Cannon, D.L. Hawksworth and M.A. Sherwood-Pike, Commonwealth Mycological Institute, 1985.

For those whose interest comes to extend beyond lichen recognition there is an excellent very short introduction to the main themes of lichenology provided by one of the Shire Natural History booklets. This is *Lichens,* by Jack R. Laundon, Shire Publications, 1986 (OOP). It brings together an extraordinary amount of essential information, with illustrations, and has proved to be very popular. Another simple expository booklet, *Lichens in Southern Woodlands,* by K. Broad, has been published for the Forestry Commission by Her Majesty's Stationery Office, 1989 (OOP). It has useful sections on lichen ecology and atmospheric pollution and also on the conservation of woodland lichens. *Lichens on Trees,* by A. Orange, British Plant Life No. 3, the National Museum of Wales, 1994 gives descriptions, colour photographs and line drawings for about 40 common species.

Detailed information on pollution effects and monitoring may be found in *Pollution Monitoring with Lichens,* by D.H.S. Richardson, Naturalists' Handbook No. 19, published by the Richmond Publishing Co. Ltd, 1992.

A more recent publication is *Lichens,* by O.L. Gilbert and published by Collins (2000) in their *The New Naturalist* series. This book gives a full account of lichens with special emphasis on habitats.

More advanced reading is provided in two student texts, one English and one American. These are:

Hawksworth, D.L. and Hill, D.E., *The Lichen-Forming Fungi,* Blackie, 1984 (OOP).

Nash, Thomas H., *Lichen Biology,* Cambridge University Press, 1996.

Both of these books deal with lichen morphology, reproduction, physiology, nutrition, ecology, air pollution and chemistry. *Lichen Biology* is, to an extent, a fourth edition of Mason E. Hale's well-known *The Biology of Lichens* but it ranges over a a wider field of topics, including recent developments in lichen classification. *The Lichen-Forming Fungi* also gives a full treatment of the essentials of lichen thallus structure, reproduction and growth.

Another useful book is *The Lichen Symbiosis* by Vernon Ahmadjian, John Wiley & Sons, Inc., 1993.

Of a more general nature there is a fascinating and very readable account of lichens in Richardson, D.H.S. *The Vanishing Lichens, their History, Biology and Importance*, David & Charles, 1975 (OOP).

Sooner or later the keen amateur will find that he or she will be impelled to read Annie Lorrain Smith's *Lichens*, 1921 (republished in 1975 by the Richmond Publishing Co.). This is a detailed treatment of all the salient features of lichenology as understood in the early years of this century and it helps greatly in the appreciation both of fundamentals and of later developments.

Important for the understanding of development of lichenology in this country is *Lichenology in the British Isles, 1568–1975*, a historical and bibliographical survey, by D.L. Hawksworth and M.D. Seaward, The Richmond Publishing Co., 1977 (OOP).

There are several collections of papers by leading scholars which, taken in sequence, explain very adequately the increasing range and depth of lichenological studies. Some of the papers they contain are daunting for all save the specialist, but a number, and in particular those referring to thallus structure, lichen symbiosis and lichen reproduction, are within the compass of a determined enthusiast. Some of these are:

Ahmadjian, Vernon and Mason L. Hale (eds.), *The Lichens*, Academic Press, 1973 (OOP).

Brown, D.H., Hawksworth, D.L. and Bailey, R.H., *Lichenology, Progress and Problems,* Academic Press, J. Cramer, 1987 (OOP).

There are three collections based on specific themes relating to various aspects of lichen ecology. These are:

Ferry, G.W., Baddeley, M.S. and Hawksworth, D.L., *Air Pollution and Lichens*, Athlone Press, 1973 (OOP).

Seaward, Mark, *Lichen Ecology*, Academic Press, 1977 (OOP).

Bates, Jeffrey A. and Farmer, Andrew M., *Bryophytes and Lichens in a Changing Environment,* Oxford University Press, 1992.

Ferry et al. contains major papers dealing with early studies in atmospheric pollution. Seaward deals with such issues as colonization, growth, succession and competition with lichen communities in hot and cold deserts and in coniferous forests, and with lichens on man-made substrata. It has also a very valuable bibliographical guide to literature dealing with lichens in different geographical regions and countries. This guide has been brought up to date and republished in *The Lichenologist* **22**(1), 1990.

A major value of these five compendia is that among their contents can be found a starting point for the exploration of both earlier literature on a chosen topic and of more recent research studies. They all contain very full lists of references.

The history of lichenology is best traced with the help of the first chapter of Smith's *Lichens*, to which mention has already been made.

Hawksworth has written a substantial and illuminating paper explaining how the foci of lichenological studies changed in the century before 1973. This is 'Some advances in the study of lichens since the time of E.M. Holmes', *Bot. J. Linn. Soc.*, **67**: 3–31, 1973.

If at all possible, reading in lichenology should not be restricted to British and American material. Indeed, some comprehensive and attractively presented studies are the work of French, German and Scandinavian scholars. Thus, in French there is the splendidly produced and illustrated work of P. Ozenda and G. Clauzade, *Les Lichens, Etude biologique et flore illustrée*, Masson, Paris, 1970 (OOP). This is a large volume of some 800 pages, with all species fully discussed and further explained through black and white drawings. It has also a very lucid and substantial introduction to lichenology in general. Also in French there is a textbook, *Guide des lichens*, by Chantal van Haluwen and Michel Lerond, Edition Lechevalier, Paris, 1993. This is distinctive, in that it presents notes indicating how various aspects of lichens (e.g. symbiosis) may be studied.

In German there is the two volume work *Die Flechten Baden-Württembergs*, by Volkmar Wirth, Eugen Ulmer, 1995. This deals solely with one region in West Germany but the lichens described are common to western Europe and beyond. Besides 555 coloured photographs of outstanding quality it contains 996 maps showing the distribution of lichen species throughout Baden-Württemberg. This is essentially an illustrated flora with keys for the region.

A guide, consisting of photographs with brief explanatory notes and dealing with ferns and mosses as well as lichens, has been compiled by Hans Martin Jahns. It is *Farne-Moose-Flechten, Mittel-Nord und Westeuropas*, Munich, BLV Verlagsgesellschaft, 1980. It has been translated into English by Mr and Mrs J.R. Laundon and published by Collins, under the title of *Ferns, Mosses and Lichens of Britain, Northern and Central Europe*, 1983. With Aino Henssen, Jahns has also produced a substantial and well-organized textbook, which covers much the same ground as both Hale and Hawksworth and Hill and which, in addition, has two sections dealing with classification and taxonomy and with the characteristics of lichen families and genera. This is *Lichenes, eine Einführung in die Flectenkunde* (Georg Thieme, 1974).

In Norwegian, but with an English supplement to the keys, there is *Lav Flora Norskbusk* by H. Krog and T. Tonsberg and published by Og Bladlav, 1980, which has excellent photographs in colour. An earlier guide to Scandinavian lichens was published in 1982. This is *Macrolichens of Denmark, Finland, Norway and Sweden*, by L. Dahl and H. Krog, published by Universitetsforlaget, Oslo (OOP). It has excellent keys to foliose and fruticose lichens, illustrated by line drawings. In Swedish a recent text, again copiously illustrated, is *Lavar: Enfälthandbok*, by Roland Moberg and lngmar Holmasen, published by Rahm and Stenstrom, 1982.

The major classic key to European lichens generally is the three volume work by J. Poelt, *Bestimnungschlussel Europäischen Flechten*, published by J. Cramer, 1969 (OOP). Poelt added two supplements with A. Vezda in 1977 and 1981. The work deals with Europe from Urals to Portugal and the Azores, including Spitzbergen, but not the Caucasus.

Mention should also be made of a massive 900 page production in Esperanto; *Likenoj de Okcidenta Europo, illustrita determinlibro*, by S. Clauzade and C. Roux, Société Botanique du Centre Ouest, Royan, France. This work is a revision and enlargement of that by Clauzade and Ozenda, listed above. It covers the whole of Western Europe from the north of Scotland and central Scandinavia, south to Spain, Sicily and Yugoslavia and from the north west of Ireland, east to Austria. The keys are well constructed and clear and it is illustrated by an enormous number of black and white drawings giving details of internal structures (asci, spores, etc.) as well as of thallus forms.

The leading journal in the field is *The Lichenologist*, published for the British Lichen Society by Academic Press. It is devoted to reports of major research and especially to those of taxonomic importance. Less substantial material, which includes occasional new keys, lists of articles and papers and reports of conferences and field excursions, is carried in the Society's Bulletin. Research papers and conference reports are to be found in a wide range of journals, including *The Bryologist* (USA), *Nova Hedwigia* (Germany), *Herzogia* (Germany), the *Revue Bryologique et Lichenologique* (France) (OOP), *Crytogamie, Mycologie* (France) and *Graphis Scripta* (Denmark).

Finally, the British Lichen Society has produced a first CD ROM, *Identification of Parmelia* Ach., British Lichen Society, 1997, with 119 colour photographs which is described as 'an essential aid to the identification of the 47 species of Parmelia, supplementing the maps and text of the first fascicle of *The Lichen Atlas of the British Isles'*.

The British Lichen Society was formed in 1958 to stimulate and advance interest in all branches of lichenology. In addition to producing a range of publications the Society organises field meetings and workshops throughout the U.K. and members have access to the Society's library and slide collection. Taxonomic help is available from Regional and Specialist referees. The Society welcomes members whatever their level of interest and experience in lichenology.

Full details of the Society's activities may be found on the Society's website at http://www.argonet.co.uk/users/jmgray or by writing for a prospectus to The Secretary, c/o The Natural History Museum, Cromwell Road, London SW7 5BD.

A very useful way to get started, or increase your knowledge of lichens is to attend a course. Courses for beginners and the more advanced lichenologist and led by experienced members of the British Lichen Society, are run by the following organisations:

The Field Studies Council, Preston Montford, Shrewsbury, Shropshire SY4 1HW
Tel: 01743 850370

Kindrogan Field Centre, Enochdhu Blairgowries, Perthshire PH10 7PG
Tel: 01250 881286

The Kingcombe Centre, Toller Porcorum, Dorchester DT2 0EQ
Tel: 01300 320684

Knuston Hall, Irchester, Wellingborough, Northants, NN29 7EU
Tel: 01933 312104 – has courses on Churchyard Lichens run by Tom Chester.

Index

Page numbers shown in bold indicate that the species is illustrated with a photograph.